Making Sense of Fi

Learn how to use field research to bring essential people-centred insights to your information design projects.

Information design is recognized as the practice of making complex data and information understandable for a particular audience, but what's often overlooked is the importance of understanding the audience themselves during the information design process. Rather than rely on intuition or assumptions, information designers need evidence gathered from real people about how they think, feel, and behave in order to inform the design of effective solutions. To do this, they need field research.

If you're unsure about field research and how it might fit into a project, this book is for you. This text presents practical, easy-to-follow instructions for planning, designing, and conducting a field study, as well as guidance for making sense of field data and translating findings into action. The selection of established methods and techniques, drawn from social sciences, anthropology, and participatory design, is geared specifically toward information design problems. Over 80 illustrations and five real-world case studies bring key principles and methods of field research to life.

Whether you are designing a family of icons or a large-scale signage system, an instruction manual or an interactive data visualization, this book will guide you through the necessary steps to ensure you are meeting people's needs.

Sheila Pontis is an information designer bridging theory and practice. She is currently Honorary Research Associate at University College London, UK, a lecturer at Princeton University, USA, teaching creativity, design thinking, and information design, and partner at Sense Information Design, a New York-area design consultancy.

Making Sense of Field Research

A Practical Guide for Information Designers

Sheila Pontis, PhD

LONDON AND NEW YORK

First published 2019
by Routledge
2 Park Square, Milton Park, Abingdon, Oxon OX14 4RN

and by Routledge
711 Third Avenue, New York, NY 10017

Routledge is an imprint of the Taylor & Francis Group, an informa business

© 2019 Sheila Pontis

The right of Sheila Pontis to be identified as author of this work has been asserted by her in accordance with sections 77 and 78 of the Copyright, Designs and Patents Act 1988.

All rights reserved. No part of this book may be reprinted or reproduced or utilised in any form or by any electronic, mechanical, or other means, now known or hereafter invented, including photocopying and recording, or in any information storage or retrieval system, without permission in writing from the publishers.

Trademark notice: Product or corporate names may be trademarks or registered trademarks, and are used only for identification and explanation without intent to infringe.

British Library Cataloguing-in-Publication Data
A catalogue record for this book is available from the British Library

Library of Congress Cataloging-in-Publication Data
Names: Pontis, Sheila, author.
Title: Making sense of field research : a practical guide for information designers / Sheila Pontis.
Description: Abingdon, Oxon ; New York, NY : Routledge, 2019.
Identifiers: LCCN 2018020031 | ISBN 9780415790024 (hardback) | ISBN 9780415790031 (pbk.) | ISBN 9781351819114 (epub) | ISBN 9781351819107 (mobi)
Subjects: LCSH: Information visualization. | Quantitative research. | Social science–Research.
Classification: LCC QA76.9.I52 P66 2019 | DDC 001.4/226--dc23
LC record available at https://lccn.loc.gov/2018020031

ISBN: 978-0-415-79002-4 (hbk)
ISBN: 978-0-415-79003-1 (pbk)
ISBN: 978-1-315-21361-3 (ebk)

Typeset in Thesis
by Sheila Pontis
Printed and bound by CPI Group (UK) Ltd, Croydon CR0 4YY

Contents

List of figures	ix
List of tables	xiii
Preface	xv
Acknowledgements	xxi

PART I: Two practices, one journey   1

1. What is information design?   3
Information design challenges	5
The role of people in information design	7
New roles, new needs, new skills	10
Emerging role: information design researchers	14

2. Bridging information design and field research   17
Understanding people-centred research in design	17
Understanding qualitative research	23
Why clients (and designers) don't trust qualitative research	24
Working with field research: the information design process revised	29
When to conduct field research in information design	34

3. What is field research?   37
Understanding key components	37
Considerations for conducting field research in information design	38
How to develop field research sensibility	44
How to ensure quality and validity in field research	44
How to work around constraints	48
Thinking creatively in field research	49
Use information design in the research process	50

PART II: Conducting a field study   53

4. How to plan and design a field study   55
    Design the study   55
    Assemble the team   68
    Put the pieces of the study together   68
    Test the design study   75
    Going into the field   76
    Checklist   80

5. Gathering data: methods for exploratory studies   81
    Observational studies   81
    Contextual interviews   89
    Contextual inquiry   95
    Design probes and diary studies   97
    Collaborative workshops   104
    Online field research   109

6. Gathering data: methods for evaluation studies   113
    Information design evaluation dimensions   113
    Assessing a design at various stages of development   115
    Assessing a design with field evaluations   118
    Concept evaluations   122
    Covert evaluations   123
    Overt evaluations   124
    Free evaluations   126

7. Making sense of field data   129
    Understanding sensemaking   129
    Sensemaking step-by-step   132
    Tools supporting analysis   145
    Methods for organizing and coding data   147
    Five Ws + One H   147
    Visual content analysis   149
    Affinity diagrams   151
    Empathy maps   154
    Methods for supporting data interpretation   157
    Needfinding   157
    Personas   159
    Visualizations   163

PART III: Communicating findings — **169**

8. Reporting field research findings — **171**
Dimensions for communicating findings — 171
Creating authentic stories — 173
How to share the whole study — 174
How to share key parts of the study — 177

9. Bridging to design: from findings to actionable design decisions — **183**
Understand findings — 184
Make findings tangible — 186
From ideas to design concepts — 192
Support the information design process — 193

10. Putting it all together — **197**

PART IV: Case studies — **201**

11. Field research in information design practice — **203**
CASE STUDY 1: The Redesign of the Carnegie Library of Pittsburgh — 205
CASE STUDY 2: Legible London — 211
CASE STUDY 3: Vendor Power Guide — 219
CASE STUDY 4: A Better A&E — 225
CASE STUDY 5: To Park or Not to Park — 231

References — **237**
Index — **245**

List of figures

1.1. Dimensions involved in effective information design — 5
1.2. Spectrum of information design problems — 8
1.3. Categories of needs based on level of generality — 11
1.4. Information design and field research relationship — 13
1.5. Field research questions across all problem types — 14
2.1. People-centred approaches used in design contexts — 18
2.2. Types of information design research — 25
2.3. Structure in qualitative methods — 25
2.4. Common views of positivists and constructive-interpretatives — 28
2.5. Research-led information design process — 32
3.1. Five key considerations for conducting field research in information design — 39
3.2. Five quality and validity criteria for field research — 45
3.3. Information design aids to support field research — 51
4.1. Field research process — 57
4.2. Mind maps — 58
4.3. Suggested methods to elicit specific types of knowledge and insights — 67
4.4. Study roadmap to plan a workshop session — 69
4.5. Study calendar — 70
4.6. Representation of materials and instruments needed for a workshop session — 71
4.7. Interview guide example — 72
4.8. Interview and observation template examples — 73
4.9. Participant log template examples — 74
4.10. Examples of different types of field notes — 77
5.1. Types of observational studies — 85
5.2. How to plan and conduct an observational study — 89
5.3. Contextual interview setup — 91
5.4. How to plan and conduct a contextual interview — 94
5.5. Contextual inquiry setup — 96
5.6. How to plan and conduct a contextual inquiry — 98
5.7. Design probes and diary studies structure — 100
5.8. Package and set of data collection instruments for a design probe study — 101
5.9. How to plan and conduct design probes or a diary study — 105
5.10. Collaborative workshop setup — 107
5.11. How to plan and conduct a collaborative workshop — 110
6.1. How to choose a field evaluation — 120
7.1. Sensemaking process for field data — 133

7.2.	Techniques to code field data	136
7.3.	Ways to label and indicate codes in the data	138
7.4.	Examples of code definitions	138
7.5.	Categories created after combining codes from interview data	141
7.6.	From codes to themes	142
7.7.	Tools for conducting field data analysis	145
7.8.	Methods for organizing and coding data	148
7.9.	Visual content analyses	150
7.10.	Visual comparative analysis	151
7.11.	Affinity diagram to make sense of interview data	153
7.12.	Empathy map to make sense of interview data	155
7.13.	Methods for supporting data interpretation	157
7.14.	Formats for creating personas	162
7.15.	Representations of concepts to visualize findings	164
7.16.	Evolution of diagrams for visualizing connections and findings	165
7.17.	Template to create a doctor and patient vaccination journey	167
7.18.	Vaccination journey	168
8.1.	Visual table to share study findings	173
8.2.	Tools for sharing key parts of the study	177
8.3.	Personal cardset template	178
8.4.	Storyboards from contextual inquiry sessions	181
9.1.	The challenge of turning findings into actionable items	183
9.2.	How to turn findings into actionable design decisions	185
9.3.	Finding cards from a contextual interview study	187
9.4.	Four quadrants template	188
9.5.	"Today and tomorrow" pictures	190
9.6.	Visual brainstorming session	192
9.7.	Use of findings at different stages of the research-led information design process	194
10.1.	Overview of field research in information design	198
11.1.	Research outputs	206
11.2.	Research outputs	207
11.3.	Analysis output	209
11.4.	Analysis output	209
11.5.	Design concepts	210
11.6.	Field research data	213
11.7.	Field research data	214
11.8.	Research session	215
11.9.	Design discussions	216
11.10.	Final prototype of the Legible London wayfinding system	218
11.11.	Vendors with the final resource	221
11.12.	Inside the resource	222

11.13.	Inside the resource	223
11.14.	Research outputs	227
11.15.	Research session	227
11.16.	Research outputs	228
11.17.	Final process map	229
11.18.	Final modular system	230
11.19.	Street parking sign in New Jersey	232
11.20.	Process sketch	233
11.21.	Old and proposed street parking signs	233
11.22.	Formative evaluation study	234
11.23.	The evolution of the proposed new street parking sign	235

List of tables

3.1.	Types of triangulation to increase credibility of findings	46
4.1.	Participants' field study details for the case studies discussed in PART IV	62
4.2.	Examples of vague and detailed field notes	79
5.1.	Overview of methods discussed in this chapter	82
5.2.	Aspects you should pay attention to when conducting an observation study	88
6.1.	Information design performance dimensions	116
7.1.	Examples of operations you can use to revise codes	139
7.2.	Questions that can help extract information from participants to create an empathy map	156
7.3.	Examples of need statements	159
7.4.	Key general and project-specific dimensions that a persona can include	161
11.1.	Overview of case studies discussed in the chapter	204

Preface

What is qualitative research? How does it differ from market research? What should I do first? Can I have a study with just five people? How can I collect data? There is not enough time to conduct a research study! What is a theme? What is coding? How does coding lead to themes? All these are frequent questions that I asked myself 12 years ago at the beginning of my career as a researcher. While I was already an experienced information designer back then, I was just taking my first steps as a qualitative design researcher. Nowadays, too often, I hear similar questions from my students and even from information design colleagues who believe that research isn't really compatible with their practice. However, the reality is that information design practice demands research more than ever, because of the many ways it is changing.

A convergence of forces—the ready availability of design technology, the Internet explosion, and the proliferation of data—has generated considerable interest in information design in recent years. The increasing complexity of today's challenges demands more frequent collaboration with professionals from other backgrounds. More and more information designers are tackling social and organizational challenges rather than only the redesign of existing solutions. As a result, the application of their skills to develop a tangible artefact as the only solution no longer plays a big role; instead, information designers must create solutions from scratch or design experiences. Paradoxically, as the need for information design skill to address complex and unframed challenges grows more urgent, the quality and performance of many information design outputs often falls short, failing to address the needs of intended audiences. Frequently, these solutions present prettier designs but with less understandable and harder to use information.

This recurring phenomenon highlights a key problem facing information design: there is too much emphasis on the production of design outputs and too little attention paid to fundamental understanding of problems and people. I have identified three main issues that contribute to this situation:

- **Working with assumptions.** Information designers tend to work with assumptions fed by previous projects. The more years of experience a team or we have, the more we prefer to start designing based on what we already know rather than to challenge any assumptions in our work. However, many of the things we think are "basic knowledge" and widely understood by our audiences, are actually very confusing to them. For example, many people have no idea how to use online banking or their smartphone; others don't understand voting instructions or how to complete a tax form[1]. This lack of understanding is particularly prevalent with technological solutions, but certainly not exclusive to them.

- **Working fast.** Some information designers have started using agile processes and design sprints as part of their practice, although the former originated for software development and the latter was developed at Google Ventures, both as ways to generate tangible solutions. The issue I see with these approaches is that they seem to perpetuate the idea that speed is everything; spending time gaining deep understanding of problems and people tends to be de-prioritized[2]. In addition, many information design problems don't involve the creation of a tangible solution.

- **Working only with user research.** Recent years have seen an increase in the adoption of research methods in design practice in general and in information design practice in particular. However, the type of research typically done takes the form of user or market research and tends to mostly focus on optimizing or evaluating a design solution. This research approach focuses on "users" not "people" and assumes that the information designer already knows enough about the intended audience, so that they can create a solution that they think the audience wants and then test how well or poorly it works[3].

Proposing a way to address these three issues and my own struggles as an information design researcher was my motivation for writing this book. Creating effective information design involves more than relying on design expertise, conducting marketing research, learning how to use specialized software[4], or working with pre-defined design briefs. Research shouldn't only be used to make sure you are designing an infographic or a website in the right way; it should be used to determine whether your intended audience needs an infographic, a website, or something else entirely. I suggest that information designers should work with a mode of research focused on gathering deeper understanding of the problem and spending time with their intended audiences. Knowledge gained from these interactions, before

1 Patel, N. (2018) Everything is too complicated: What are you assuming people already know? *The Verge* [online], Available at: https://www.theverge.com/2018/1/7/16861056/ces-2018-bad-assumptions-smart-assistants-tech-confusion [Accessed 9 January 2018].
2 Hall, E. (2017) Design Sprints are Snake Oil [online], Available at: https://medium.com/research-things/design-sprints-are-snake-oil-fd6f8e385a27 [Accessed 9 January 2018].
3 Roberts, S. (2017) The UX-ification of research, *Stripe Partners* [online], Available at: http://www.stripepartners.com/the-ux-ification-of-research/ [Accessed 5 January 2018].
4 Heller, S. & Landers, R. (2014) *Infographic Designers' Sketchbooks*. New York: Princeton Architectural Press.

deciding what the solution should be, will help build empathy and inform decisions throughout the design process that will lead to higher-quality solutions. *Field research* is the type of qualitative research that provides this level of insight and helps reveal the complexity of everyday life; it isn't simply a complementary approach to marketing research. As I see it, field research is about the unknown, unexpected, and unanticipated.

As information designer David Sless pointed out back in 2008[5], the combination of qualitative research with design knowledge, principles, activities, and processes isn't new; it has a long, but often marginal, history in information design, with a strong presence only in particular places (e.g. the UK) and larger design studios and organizations (e.g. Applied Wayfinding, Design Council UK). Nevertheless, the use of field research in information design isn't a common practice, and, in general, the response to qualitative research remains hostile:

- 'This isn't the designers' job'.
- 'Time frames are too short'.
- 'There aren't enough resources'.
- 'Directing a conversation and listening at the same time is too hard'.
- 'No one will talk to you for more than ten minutes'.
- 'I don't need research, personas can be made up from our imagination'.
- 'I have been an information designer for 20 years, I know what my audience needs'[6].

Rather than attempt to go into the field and try to use qualitative research, information designers cling to these and other myths and misconceptions, preventing them from developing confidence, deep understanding, and patience needed to use this type of research in their practice. Conducting research that involves actual contact with intended audiences seems intimidating, time-consuming, and expensive.

I believe that there is no better time than the present to break from these myths, get immersed, and gain deeper awareness about this form of research. The need to understand people's contexts and behaviours has become a non-negotiable step in the information design process to create successful solutions. This book advocates the use of the investigative rigour and systematic methodologies of field research in combination with the dis-

---

5 Sless, D. (2008) Measuring information design, *Information Design Journal*, 16(3), 250–258.
6 Students, teachers, colleagues, and clients from various countries (e.g. Argentina, Spain, Switzerland, Finland, Portugal, Germany, the Netherlands, the UK, the US) have expressed these myths and assumptions during the last decade.

ciplined logic and visual principles of information design. In field research, methods are used to study and understand cultures and communities through observation and interpretation of traits and people's behaviours in their environments. The selection of methods discussed here comes from context mapping, contextual inquiry, human sciences, and human-centred and participatory design. For most methods, the number of participants is small (6–20). These methods are relevant to professional contexts, in that, used in the field, they provide a deeper view into participants' lives and, in some cases, give participants the opportunity to immerse themselves in the information design process.

This book presents both field research and information design as complementary practices. The former follows more established methodologies to understand people's lifestyles, behaviours, needs, and emotions, where its practitioners are well equipped to understand cultures, but they don't have the skills to visually make sense of and clearly communicate complex information and findings. The latter aims to help people make sense of situations and uses a more visual set of skills to identify patterns and analyse and communicate information, but its practitioners are less prepared to observe the field. While field research offers a way to observe and get in touch with people's complexity, information design offers a way to make sense of and communicate that complexity. These combined sets of skills are needed to successfully tackle the complexity of current challenges. I propose the use of information design skills to tailor field research methods and create tools for the specifics of each project. This means that a same tool or method could be used at different steps for different purposes.

The goal of this book is to provide more robust support and clear guidance on both deciding how to gather field data to help inform and move forward in the information design process, and analysing and making sense of the collected data. The aim is to provide direction primarily to professionals, students, and researchers associated with information design, with a secondary audience of students and professionals from the social sciences, anthropology, and psychology interested in learning visual skills to improve their work and create more compelling stories. Suggestions and tips aren't meant to impose a unique way to conduct field research but to indicate which aspects from the research process are essential to make it more credible, reliable, and efficient.

Most research books target graduate students or post docs working on their dissertations, and the language is mainly academic. This book is written for you: the information design practitioner or undergraduate student who isn't familiar with research jargon and terminologies. The idea of writing a book specifically for information designers began while I was doing my PhD, when I started conducting research to help me answer my research question. Back then, I had to learn the basics from various human sciences books and then extrapolate concepts and methods to address design needs. Since then, many books have been published to help designers gain the basic understanding of what research involves and how it can benefit design practice. However, few books explicitly discuss the use of field methods in information design or do so in plain language, understandable to designers. Furthermore, while there are books presenting a wide range of methods for gathering human-centred data, the subsequent phases, such as analysing, making sense of, inferring, communicating, and using it in respective steps of the information design process, have received less detailed attention. That is, how to get value from qualitative data. These steps are crucial for the development of successful solutions, and it is often where design professionals struggle the most. Although the analysis and sensemaking of large amounts of information are core tasks of information designers, they don't often realize how to apply these skills in the research process. This makes the narrative of this book necessarily neither academic nor exhaustive. Rather, it has a practical focus, illustrating concepts through visuals and indicating further readings, where you can deepen your understanding of specific aspects of field research. Overall, research conducted in professional contexts has different standards from that of academic studies because it has a different end goal. The book provides guidance for conducting applied field research and doesn't discuss academic standards.

An important part of the writing process has involved doing my own field research to gain a better understanding of how information designers work and the role that qualitative research plays in their practice. I conducted in-depth interviews with some of the information designers involved in the case studies, as well as clients, to understand how they see these methods contributing to a project. These interviews have been extremely illuminating, helping to uncover common research-related challenges that information designers face with clients, and issues frequently experienced when planning and designing a field study. Where relevant, I share excerpts from these.

The book is organized in four parts:

- **PART I** introduces a working definition of information design and applied field research. It provides examples of current information design challenges and describes how the role of information designers has changed. It presents an information design process, enhanced by the use of field research, which then becomes the backbone for the rest of the book.

- **PART II** provides a roadmap for planning and designing a field research study and gives guidance for going into the field. It describes methods to gather data and evaluate solutions in context, offering direction on which methods to use. Also discussed in this part are the use of visualizations to support the research process and a step-by-step guide to make sense of field data.

- **PART III** explains ways and techniques to communicate findings to the rest of the team, intended audiences, and clients, and how they can be used at different moments of the information design process.

- **PART IV** describes the use of field research in information design practice. Five case studies explain how this approach was essential to the development of successful solutions. The selection of cases has been made from the analysis of 30 case studies from a range of information design fields and years of experience working with the methods in professional contexts.

This book is intended to provide practical guidance, tools, and techniques for conducting high-quality field research in information design practice. My hope is that it helps you to better understand the value of this type of research for your practice and to confidently conduct a field study as part of your next information design challenge.

Sheila Pontis
March 2018

Acknowledgements

The seed of this book was planted more than ten years ago when I started my own journey as an information design researcher. Since then, too many people have directly and indirectly contributed to this final form. The book also builds on my previous research work and design practice, and has been nurtured by the many students, workshop participants, and colleagues that I have encountered throughout this journey. Many thanks!

Many thanks to Kristina Abbotts, my editor at Routledge, and her team for the encouragement and support along the way. Linda March proofread the manuscript. Tim Fendley, Ben Acornley, and Savannah Kuchera from Applied Wayfinding took the time to share the process behind the scenes of Legible London and gave permission to include images from the project. Many thanks also to Tom Lloyd, Danai Papadimitriou, and Kelly Pollard from PearsonLloyd, David Bishop and Bridget Deely from MAYA, Nicole Sylianteng, Candy Chang, and Jazlyn Patricio-Archer from the Center of Urban Pedagogy, for also giving permission to include images from your respective projects.

Special thanks to Prof. Ann Blandford for your wisdom and inspiration, and providing comments on an earlier version of the manuscript.

Finally, my gratitude to my family: Graciela Salerno and Horacio Pontis for always being there, asking questions, and trying to understand a very different field from yours. And to Michael Babwahsingh for reading many earlier versions of the manuscript, and for your endless patience and support. Our many information design conversations have also inspired this book, challenging my thinking and helping me articulate my ideas with more clarity.

PART I:
Two practices, one journey

1 What is information design?

The information design community actively writes and theorizes about various aspects of the field (e.g. methods, tools, skills, processes), and case studies abound, but there is a lack of clarity regarding what information design is or what its boundaries are, and an agreed definition of the field is still elusive[1]. Perhaps this is the result of the diversity of disciplines that inform the field, including graphic design, journalism, interface and user experience design, cognitive science, behavioural and applied psychology, and information science, among others[2]. This cross-disciplinary and multi-faceted nature of information design makes it challenging to arrive at a concise definition that accurately captures its breadth and depth[3]. On the other hand, the influence of these disciplines has helped to better equip information designers with a rich toolkit of skills. Terry Irwin[4] summarizes information designers as:

'...very special people who must muster all of the skills and talents of a designer, combine [them] with the rigor and problem-solving ability of a scientist or mathematician and bring the curiosity, research skills and doggedness of a scholar to their work.'

This book broadly defines information design as the field concerned with facilitating understanding in order to help people achieve their goals by translating raw or disorganized data into forms that can be rapidly perceived, understood, processed, and used. Information design work of any kind seeks to enhance understanding—of a situation, concept, space, place, time, quantity, phenomenon—for an intended audience[5]. Information designers aim to design clear communication on any medium from paper to digital devices and public information displays[6]. Regardless of context or project type, their goal is to maximize benefit and value for the client and end-user.

---

1 Jacobson, R. (2000) *Information Design*, London & Cambridge, MA: The MIT Press.
2 Waller, R. (2011) *Technical Paper 14, Information design: How the disciplines work together*, Simplification Centre, University of Reading.
3 Gobert, I. & van Looveren, J. (2014) *Thoughts on Designing Information*, Zurich: Lars Müller Publishers.
4 Irwin, T., 2002. Information design: What is it and who does it? *American Institute of Graphic Arts (AIGA)* [online], 21 June, Available at: http://online.sfsu.edu/jkv4edu/2DMG/projects/Informationdesign.pdf [Accessed 14 November 2017].
5 Wurman, R.S. (1996) *Information Architects*, Zurich: Graphics Press.
6 Pettersson, R. (2010) Information design – Principles and guidelines, *Journal of Visual Literacy*, 29(2), 167–182.

When we encounter raw data or unstructured information (e.g. highly specialized, confusing, or large amounts), we may experience anxiety, because our brain can't identify patterns or connections or process what we see. As a result, we perceive raw and unstructured situations as **complexity**[7]. To help people process situations and make connections, information designers give structure to raw hard or soft data, designing it in a way that makes it more accessible and meaningful. Most design decisions made throughout this process aim to support the intended audience's **cognitive activities** and **sensemaking**.

Sensemaking is the process of understanding something (a situation or problem) which occurs when there is an intentional effort to discover connections among data (e.g. people, places, events) and find meaning from information[8]. Through this process, we engage in cognitive activities (e.g. memory, perception, visual processing) to filter relevant from irrelevant information, give meaning to experiences, move from data to an interpretation, and construct broader understanding of a specific situation. As cognitive artefacts, information design outputs support and strengthen these mental abilities and functions, allowing the brain to process information more effectively.

When **information is effectively designed**, it facilitates navigation, supports collaboration, improves usability, and enables understanding. Effective information design involves a deep understanding of four dimensions (Figure 1.1):

- **Problem:** what challenge the audience encounters
- **People:** who the audience is and their needs
- **Context:** how, where, and when the audience accesses the information
- **Content:** what needs to be communicated

Information designers can help wherever there is a need to, for example, make decisions, build a strategy, understand and solve a problem, create a presentation, or write a report. This need for understanding is discipline-agnostic, spanning all disciplines and industries.

---

7 Wurman, R.S. (1989) *Information Anxiety*, New York, Doubleday.
8 Weick, K.E. (1995) *Sensemaking in Organizations*, Thousand Oaks, CA: SAGE; Klein, G., Phillips, J.K., Rall, E.L., & Peluso, D.A. (2007) A data-frame theory of sensemaking, in *Expertise Out of Context, Proceedings of the Sixth International Conference on Naturalistic Decision Making*, Erlbaum, 113–155.

DIMENSIONS INVOLVED IN EFFECTIVE INFORMATION DESIGN

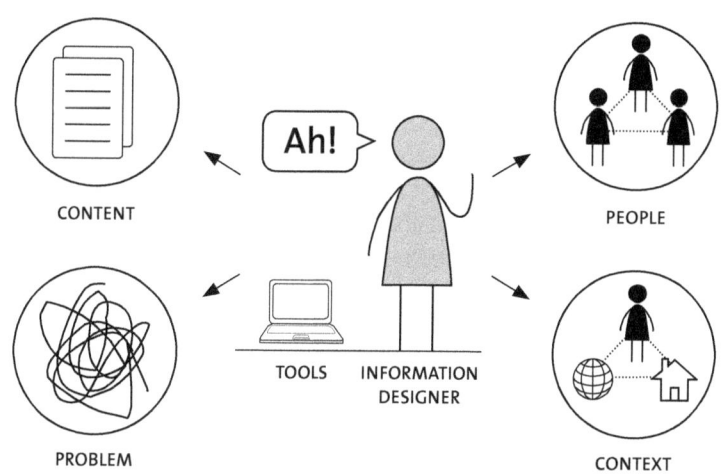

Figure 1.1
In addition to skills and tools, effective information design involves the understanding of four dimensions: Problem, People, Context, and Content.

Information design challenges

The domain of information design challenges is broad and diverse and isn't only limited to the complex; even the simplest concept can be miscommunicated and requires no less rigour and attention to be conveyed clearly[9]. Today, information design challenges come in various shapes and sizes, reflecting the richness of everyday interactions. Some information designers work in diverse projects from various areas, while others specialize on one or two areas. The following are the most common information design specializations: 1) icons and symbols to aid identification, 2) visual explanations to show how things work or how to do something (e.g. information graphics, instructions), 3) data visualizations to show quantities (e.g. pie charts, line graphs), 4) documents to manage and access information (e.g. tax forms, gas bills), 5) interactions to help access dynamic information environments (e.g.

---

9 Siegel, A. & Etzkorn, I. (2013) *Simple: Conquering the Crisis of Complexity*, London: Twelve; Pontis, S. & Babwahsingh, M. (2013) Communicating complexity and simplicity: Rediscovering the fundamentals of information design, *2CO COmmunicating COmplexity*, Alghero, Sardinia, Italy, 25–26 October 2013, 244–261; Frascara, J. (ed.) (2015) *Information Design as Principled Action: Making Information Accessible, Relevant, Understandable, and Usable*, Champaign, Il: Common Ground Publishing.

websites, apps), 6) maps and signage systems to navigate spaces, and 7) systems to show how parts are connected (e.g. strategies).

At the same time, information design principles and skills are increasingly necessary in a wide range of domains. In these cases, information design skills help externalize thoughts and ideas, promote human-centredness, encourage systems thinking, facilitate cross-disciplinary collaboration, and also support sensemaking. These challenges and needs are found in domains like science and technology, pedagogy, urbanism, or arts and humanities, among others. To better understand the rich spectrum of challenges in which information designers are directly or indirectly involved, let us examine two broad types of challenges they frequently encounter[10]:

- **Well-defined or framed problems** are highly framed and clearly stated, with a readily conceivable or tangible solution, such as an app, an information graphic, a wayfinding sign, or a bus timetable. The problem is clearly defined and the solution known. Well-defined challenges often integrate a large proportion of information design skills—for example, signage design skills are needed for wayfining projects or editorial design skills are needed for document design projects—and a smaller proportion of other specialized sets of skills. The vast majority of information design projects, including the redesign of existing products or solutions, fall into this type because the problem statement suggests what could be designed. Other challenges of this type are those in which the problem is also explicitly defined but the solution is unclear. Unlike the first type, solving this type of challenge requires additional learning, in that previous experience can't be solely relied upon, and may involve a larger proportion of non-information design skills.

- **Ill-defined or unframed problems** lack distinct boundaries and require greater effort to frame and address because they are, by nature, ambiguous, often involving close interaction with individuals from other disciplines. Both the problem and the solution are undefined, and something new should be created from scratch. Information designers' role in this context is less about providing direct solutions than making aspects of the problem clearer and more manageable, as well as increasing understanding of how something works and giving visual form to envisioned goals and outcomes. To understand and solve such problems, extensive learning is required; diverse and cross-disciplinary teams should work together, and frequent input and collaboration from stakeholders and audience is necessary. In this type of challenge, the proportion of information design skills changes, as other sets of

---

10 Conway, R., Masters, J., & Thorold, J. (2017) *From Design Thinking to Systems Change*, RSA, Action and Research Centre.

skills are also required to achieve successful solutions. Organizational change, social innovation, environmental, and systemic challenges are examples of this type of problem.

Projects and work described in the latest publications, *Information Design as Principled Action*[11] and *Information Design: Research and Practice*[12], already indicate this evolving role of the field beyond artefacts and towards the design of systems, strategies, and experiences. Figure 1.2 shows examples of each type of problem. This diversity of challenge types demands that information designers work in a more systematic and rigorous way, using different tools and methods to help them identify the dimensions of the problem at hand.

The role of people in information design

Information design is a *people-centred* field. To some extent, this maxim seems redundant because the very goal of the field is to design information in a way that it is usable and understandable to people—a specific audience—regardless of the type of challenge. People should always be at the centre of information design work and, for a solution to succeed, a deep understanding of their needs should inform decisions made throughout the information design process.

However, this isn't always the case. In practice, people's needs and limitations aren't always thoroughly considered, and decisions aren't anchored on objective evidence. The more framed a challenge is, the more commonly this scenario manifests itself, because information designers tend to rely more on their prior experience than on gathering fresh perspectives from the intended audience. This way of working is harder to sustain when dealing with unframed challenges because not even seasoned information designers may have previously worked on a similar project or the project's complexity may be so great that first-hand insights from the intended audience are necessary to frame it.

When a problem is highly ambiguous and defining its boundaries is specially challenging, information designers aren't fully prepared; in some cases, they can even struggle to know where to start or what tools, methods, or skills to use. In these cases, personal experience and knowledge alone aren't enough to create effective solutions. A more holistic type of understanding

---

11 Frascara, J. (ed.) (2015) *Information Design as Principled Action: Making Information Accessible, Relevant, Understandable, and Usable*, Champaign, Il: Common Ground Publishing.
12 Black, A., Luna, P., & Lund, O. (Eds.) (2017) *Information Design: Research and Practice*, London: Routledge.

SPECTRUM OF INFORMATION DESIGN PROBLEMS

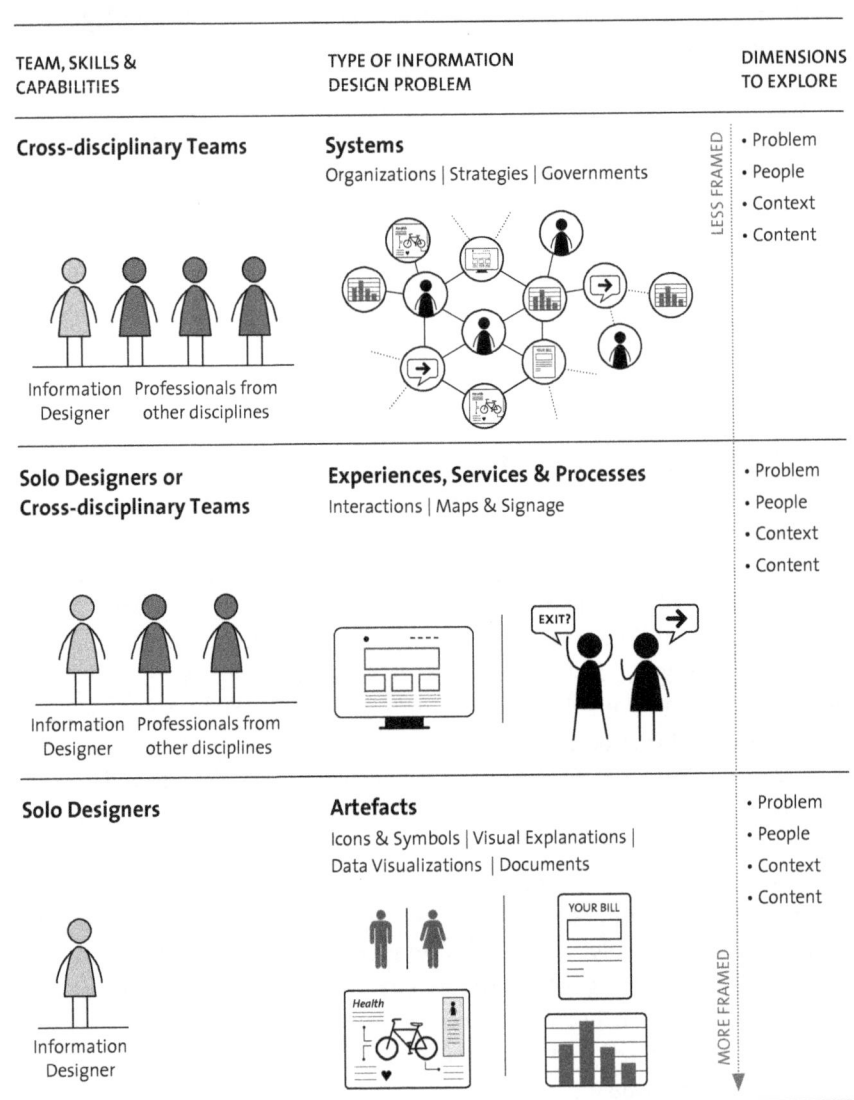

Figure 1.2

Spectrum of information design problems with examples of possible solutions, organized from more framed, such as the design of artefacts, to less framed involving the design of systems. Regardless of how framed a problem is, information designers should explore the same four dimensions.

is needed of who is involved in those situations, as well as specific learning about the intended audience's lifestyles, their experiences, and environments. Even when working on more framed projects, like the design of an information graphic, this level of understanding is necessary to avoid making design decisions mostly on assumptions. This indicates that information designers should focus on eliciting a more accurate picture of people's lives.

To gather this type of knowledge and improve the quality of work, many information designers have already added some form of *research* to their toolkit. Mostly, this involves *market research* or research conducted in a "lab" environment, through the use of questionnaires, surveys or focus groups, or the design of usability evaluations or before/after studies[13]. These types of research help identify general needs or preferences, pinpoint usability problems or measure performance. However, findings don't always provide accurate indicators because data is collected away from the intended audience's environment or the context of use (e.g. crowded and fast-paced space, with visual noise from other stimuli). For example, some information design studios test the effectiveness of city maps by inviting target users to in-house workshops or sessions to observe how they use the map. Participants' feedback on this experience may not accurately reflect their everyday reactions when using the map. In reality, they would interact with the map in a different frame of mind (e.g. commuting to work during rush hour or engaging with their phones). In addition, this type of research doesn't explore cultural trends or observe people's behaviour in everyday practices.

However, to find the specifics and nuances relevant for each project, exploring these dimensions is essential for information design practice. A more immersive approach to understanding people can provide a more holistic view of their values, attitudes, needs and struggles. Similarly, Tim Fendley[14], partner and creative director of Applied Wayfinding (London, UK), believes that information designers bring a subjective element to each project, delivered by their design experience. Research helps deliver an objective element or view to guide and give direction to decision-making. He adds that, without a deeper understanding of how people interact in and with their natural environments, design decisions will be made only on subjective elements; that is, your experience.

Gaining a deeper understanding of people's lifestyle and needs helps develop an *empathic focus*. Designing with empathy means making decisions based on what the audience needs to make life easier, rather than what you may think would work best for them. In many cases, this is easier said than done. Often, the rapid pace of information design practice pushes information

---

13 Sless, D. (2008) Measuring information design, *Information Design Journal*, 16(3), 250–258.
14 Telephone interview conducted with Tim Fendley and Ben Acornley, partners and creative directors at Applied Wayfinding (London, UK), 17 August 2017.

designers to act fast and make quick decisions. In these situations, objective evidence from research studies helps designers make decisions with confidence.

One way to minimize subjective design is to understand those things that intended audiences are missing: these are their *needs*. There are four categories of needs based on their *level of generality*[15]: common, context, activity, and qualifier needs; that is based on whether a need may apply to every single person or just to a particular sample or segment (Figure 1.3). Needs represent problems that are related to current situations, solutions, and behaviours. The most universal needs indicate deeper problems, harder to address with a single solution. Information design focuses on addressing context and activity needs, or improving an existing solution to address a qualifier need. The identification of these types of needs earlier in a project can help information designers anchor design decisions and provide a roadmap for the creation of a solution that is likely to support people's cognitive activities. For this, information designers need to work with a different type of research, such as the one introduced in the next section.

New roles, new needs, new skills

Changes in the information design field discussed throughout this chapter have also generated a change in the tasks information designers perform and skills they need. As Rubin and Chisnell (2008) put it a few years ago, 'No longer can design be the province of one person or even of one specialty.'

Information designers have long moved away from being disconnected professionals working alone to being part of large, cross-disciplinary teams of professionals working with or within bigger organizations and, in some cases, closer to the intended audience. More and more, teams are composed of specialists from many disciplines, including engineering, marketing, social science, user-interface design, human factors, and multimedia. In turn, to facilitate dialogue and develop effective cross-disciplinary work, many of these professionals have also started undertaking training in design areas[16]. Similarly, information designers are learning new skills and speaking other "languages" so they are prepared to wear more than one "hat" when necessary.

In this context, the role of information designers has expanded, and their tools have an impact beyond the boundaries of design studios. They have evolved from being only problem solvers and creators of artefacts to also being problem finders, and facilitators of dialogue, collaboration, and

---

15 Patnaik, D. (2014) *Needfinding: Design Research and Planning*, Amazon.
16 Sanders, E.B.N. & Stappers, P.J. (2008) Co-creation and the new landscapes of design, *International Journal of CoCreation in Design and the Arts*, 4(1), 5–18.

CATEGORIES OF NEEDS BASED ON LEVEL OF GENERALITY

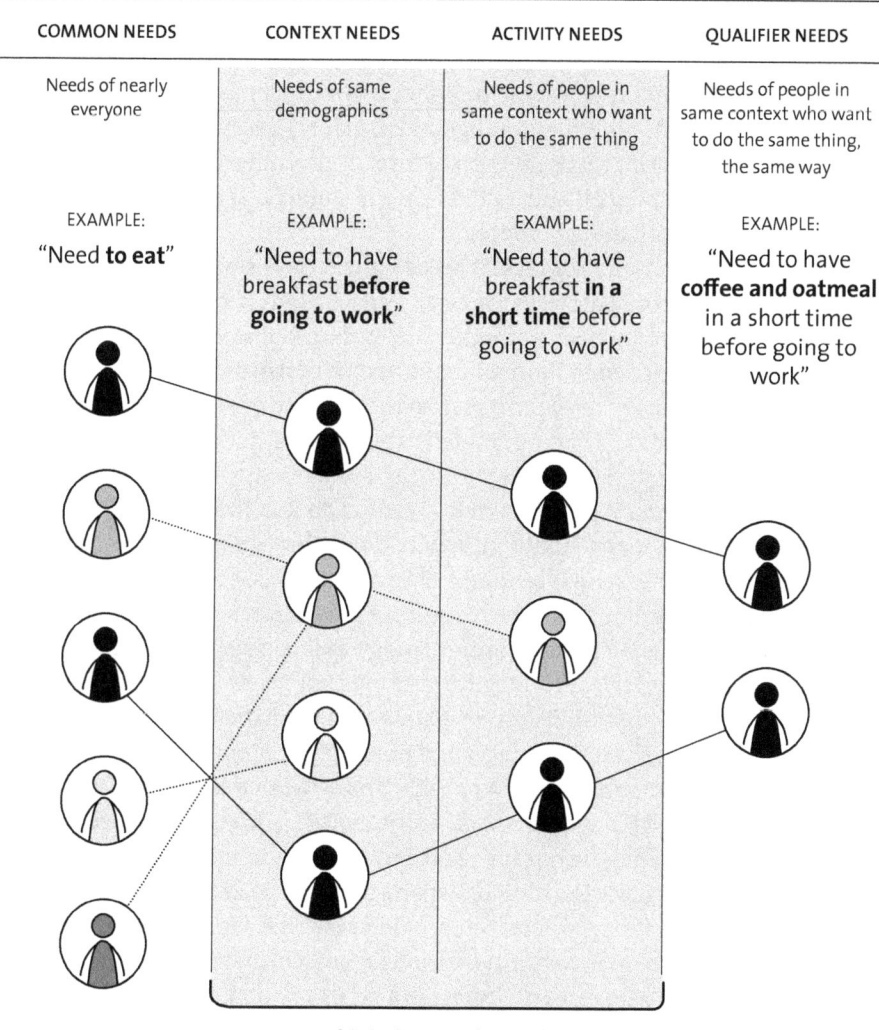

Figure 1.3
People can explicitly articulate activity and qualifier needs because they represent problems with current solutions. Information designers can address these needs with the design of new solutions. Context needs are harder to articulate because they are related to the situation in which people are and their goals. Once identified, information designers can design ways to change the situation to address these types of needs.

What is information design? | 11

understanding. Furthermore, information design techniques, such as mind mapping, colour-coding, systems thinking, visualization, and large-scale drawing, are also helping communication and enabling focus on large-scale problem-solving by revealing patterns and connections within vast datasets, from which it would otherwise be difficult to extract meaning. As facilitators, information designers encourage stakeholders to verbalize ideas and concerns, while they put them into visual form and synthesize them into a common picture, helping all parties to be on the same page as they share a common language and understanding.

As previously mentioned, information designers have started adding research studies as part of a project's life cycle. Furthermore, in many cases, the information designer and the researcher are, in fact, the same person[17] . To some extent, this has opened the door to the use of information design skills to support the analysis of research data and the creation of visualizations to communicate insights[18] . However, there is still work to do for information designers to fully adopt research as part of their practice as within the information design community, there remains confusion and little understanding of what research really entails or how to do it. This confusion increases when other types of research beyond surveys or usability evaluations are discussed, such as the one this book concerns: **field research**.

Field research[19] has its roots in anthropology and sociology, focusing on eliciting people's physical, psychological, and cultural needs. This type of research can help information designers gain a more holistic understanding of all dimensions involved in a project and how they relate to each other by identifying who is important and who isn't, and why; understanding their behaviours and attitudes in different environments (e.g. likelihood to use an app and how they would use the app); and determining differences between groups (e.g. specific characteristics for experts and for novices)[20] (Figure 1.4).

Rather than a specific type of research, field research is more an approach for understanding people, involving various research methods and techniques. In this book, "field research" refers to a group of qualitative, contextual research methods that can be used to achieve this goal. The next chapters describe this approach in detail and how information designers can use it.

17 Visser, F.S., Stappers, P.J., Van der Lugt, R., & Sanders, E.B.N. (2005) Contextmapping: Experiences from practice, *CoDesign: International Journal of CoCreation in Design and the Arts*, 1(2), 119–149.
18 Miles et al. (2013) and Beyer and Holtzblatt (2014) stress the role of communication design as an essential skill to ensure good communication of data and insights. Miles, M.B., Huberman, A.M., & Saldaña, J. (2013) *Qualitative Data Analysis*, SAGE; Beyer, H. & Holtzblatt, K. (1998) *Contextual Design: Defining Customer-Centered Systems*, San Francisco: Elsevier (https://www.interaction-design.org/literature/book/the-encyclopedia-of-human-computer-interaction-2nd-ed/contextual-design) and (2014) *Contextual Design: Evolved*.
19 Frascara, J. (ed.) (2015) *Information Design as Principled Action: Making Information Accessible, Relevant, Understandable, and Usable*, Illinois, Champaign, Il: Common Ground Publishing.
20 Chipchase, J. (2017) *The Field Study Handbook*, 2nd edn, Field Institute.

INFORMATION DESIGN AND FIELD RESEARCH RELATIONSHIP

Today

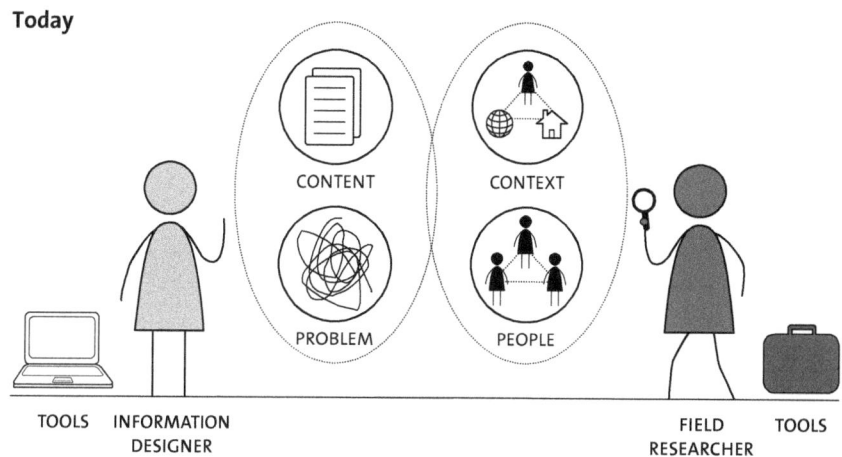

Today, information designers often focus on each dimension at a time, primarily seeking understanding of *Content* and *Problem*. Field researchers mostly focus on understanding *People* and how *Context* impacts their lives.

Tomorrow

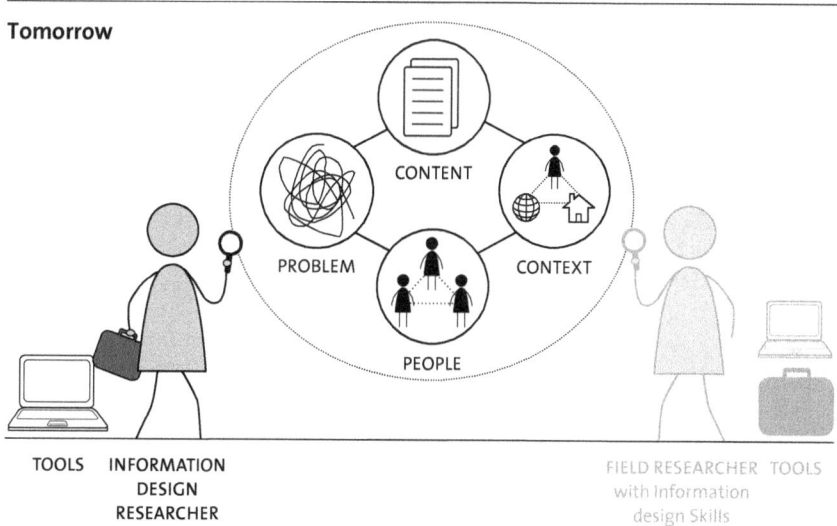

By adopting field research as part of their practice, information designers can work with a more holistic approach to understand all dimensions, how they connect, and how they affect each other. As this book primarily focuses on information designers, the role of the field researcher with information design skills isn't discussed further.

Figure 1.4
Comparison between current and desired state of the relationship between information design and field research.

What is information design? | 13

FIELD RESEARCH QUESTIONS ACROSS ALL PROBLEM TYPES

	PROBLEM	PEOPLE
Systems	• What is the scope or extent of the problem? • Why does this situation exist? • What are the dimensions of the problem?	• Who are all the different stakeholders involved? • What are their attitudes?
Experiences, Services & Processes		
Interactions	• Why does this experience need to be improved? • What other problems stem from this?	• How do people use the new design? • Does the use of the design help people achieve their goal? Why? Why not?
Maps & Signage	• What do people struggle with when navigating the space? • Why do people get lost in the city?	• What do people need to know to get from A to B? • How do people use the new map? • How do people conceptualize the space?
Artefacts		
Icons & Symbols	• Is the intended message coming across? • What unintended messages are the icons communicating?	• What is my audience's visual literacy? • What do people need to support their understanding? • What colours are more helpful for my audience?
Visual Explanations	• Why is the topic hard for people to understand? • Do people need an infographic?	• How familiar are people with the subject? • What types of explanation resonate with people? • What formats do people prefer?
Data Visualizations	• What questions need to be answered with the data visualization? • Is the data visualization communicating clearly?	• Can people understand the main topic after reading the visualization? • How familiar are people with different forms of data visualization?
Documents	• Why can't people complete the form correctly? • What aspects of the document are unclear?	• What is people's attitude towards the current document? • How does the user role change the function of the document?

Emerging role: information design researchers

Today, to accomplish high-standard solutions that improve understanding and set good communication practices, it is essential to combine the best of both worlds: research and practice. Information designers need not become full-time researchers or replace creative thinking with a solely research-based practice, but they can add field research to the information design process. As an example, Figure 1.5 presents representative questions

CONTEXT	CONTENT
• Where does the problem occur? • What are the dynamics between the different stakeholders in this situation? • In what domains does the situation exist (political, economical, social, organizational)?	• What policies or procedures are in place regarding the situation? • What do people know about the broader situation and its components?
• How does the environment impact the experience? • How does the channel affect the interactions (digital or physical)?	• What information could improve people's experience? • What terminology would be appropriate?
• How does the physical environment affect existing signage? • Which places are more crowded?	• What other information would help people navigate the space? • How much guidance is enough?
• How are icons understood in this cultural context? • How does the physical environment affect legibility of icons?	• What are all possible meanings of this icon/symbol? • What historical precedents may affect the understanding and use of this icon/symbol?
• Where would people see the information? • How does the format affect people's understanding (digital or physical)?	• What type of content should be included or left out of the infographic? • How should the information be organized?
• Where is the data visualization used? Static or interactive? • What else needs to accompany the data visualization?	• How much data should the visualization show? • Which graphic form communicates clearest?
• Where do people use the document? • How does location (home, work) affect use of the document?	• What is required and what is flexible or optional? • How many languages does the document need to be in? • What aspects of the document do people like/dislike?

Figure 1.5

In addition to answering general questions about who the intended audience is, what they need, what their behaviours are, or how much they know about a specific problem, field research can help answer more specific questions in each of the four dimensions and for each type of information design problem.

for each specialization that information designers can answer using this approach. Both **better understanding of qualitative research** and, particularly, the **use of visualizations throughout the research process** can close the gap between information design and research, constituting a step towards the adoption of this type of research in professional practice.

This book proposes a way to bridge this gap by providing information designers with guidance and methods to design and conduct field research studies as part of their practice[21]. This knowledge expands the information design skill set by adding complementary capabilities to expertise. The challenge for information designers is to integrate their creative and apparently "messy" way of working with a more systematic and rigorous approach, focused on collecting evidence from people in the field and using that to enhance and support the information design process and decisions made along the way[22].

The division between research and practice won't go away, but one step towards closing that gap is to have information designers and researchers learn how to work together and share skills. Both can contribute to each other's practices, as well as learning from each other. Having a closer relationship, more fluent dialogue, and seeing each other's skills as complementary will help both to tackle problems more successfully.

The first step to achieve this goal is to better understand the relevance of field research for information design. The next chapter starts focusing on this by examining how designers (not only information designers) have studied people throughout the years.

---

21 Proportionally, in information design, more research is still conducted in academic settings by scholars investigating different information design dimensions than in practice by professionals working on client projects.

22 Sless, D. (2008) Measuring information design, *Information Design Journal*, 16(3), 250–258.

2 Bridging information design and field research

The importance, to the design community, of studying people was formally introduced during the 20th century, after the rational wave of design and methods movement that characterized the 1960s[1]. Various research approaches from the social sciences, anthropology, psychology, ergonomics, human factors, and other human sciences made their way to the practice of design, shaping the approaches used today. From different angles, each approach gives designers tools to gain an understanding of their intended audiences[2].

Understanding people-centred research in design

Particularly, five approaches—ethnography, contextual inquiry, usability testing, participatory design, and self-documentation—have been relevant for information design (Figure 2.1). History isn't linear nor happens in a vacuum, and it is hard to describe each approach with just a few milestones, without viewing the bigger picture of what was occurring in other parts of the world, but the following overview presents a chronological storyline of how designers have used these approaches to better understand their audiences.

Ethnography

Ethnography originated in social research and anthropology to explore the nature of a particular social phenomenon, rather than testing out hypotheses about it. This "reflective process" involves first-hand empirical investigation through long periods of detailed observation of people in their environments. Traditionally, ethnographers' primary method has been *participant observation* (Chapter 5) through intensive fieldwork, where they observe people's lives, actions, and changes over time, while listening to what is said and informally asking questions[3]. The observer takes an explicit role in the

---

1 Conley, C. (2004) Where are the design methodologists? *Visible Language*, 38(2), 196–215; Jones, C.J. (1992) *Design Methods*, 2nd edn, New York: John Wiley.
2 Sanders, E.B.N. (2002) From user-centered to participatory design approaches, Chapter 1 in Frascara, J. (ed.) *Design and the Social Sciences: Making Connections*, London: Routledge.
3 Sangasubana, N. (2011) How to conduct ethnographic research, *The Qualitative Report*, 16(2), 567–573.

PEOPLE-CENTRED APPROACHES USED IN DESIGN CONTEXTS

ETHNOGRAPHY
- Observations

CONTEXTUAL INQUIRY
- Observations
- Contextual Interviews

PEOPLE IN CONTEXT

PARTICIPATORY ACTION RESEARCH
- Collaborative Workshops
- Observations
- Contextual Interviews

SELF-DOCUMENTATION
- Diary Studies
- Design Probes
- Cultural Probes

USABILITY TESTING
- Observations
- Task Analysis
- Contextual Interviews

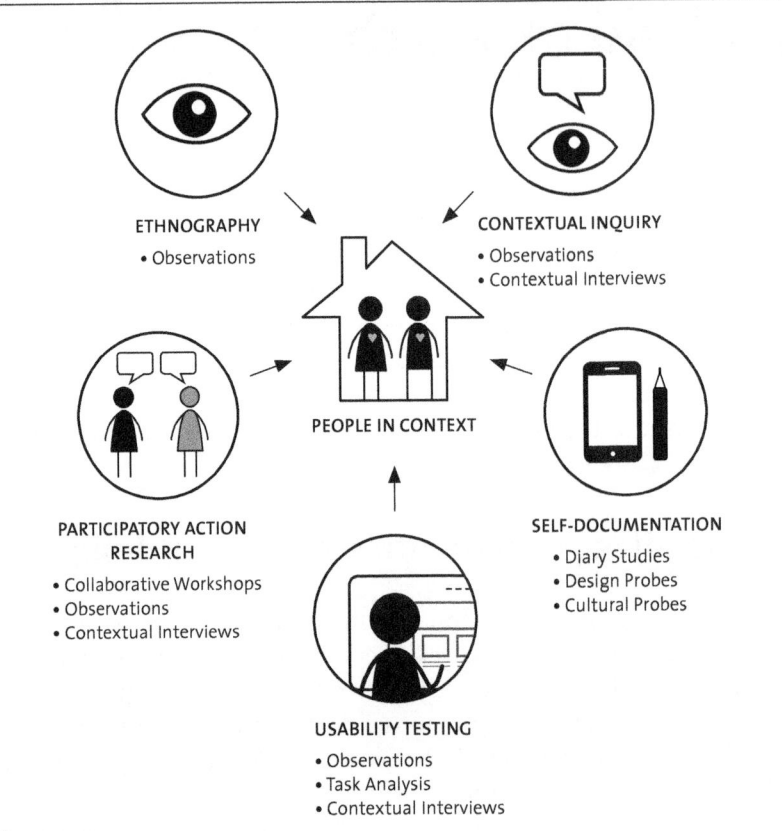

Figure 2.1
People-centred approaches and supporting methods frequently used in design contexts to better understand audiences. Usability testing is often done in labs rather than in intended audiences' natural environments.

observed environment—e.g. as a trainee or performing one of the simpler roles in that environment.

While, initially, ethnography was used to study remote cultures and people, this approach has undergone several transformations. The first was in the research focus, which moved from studying foreign cultures to a narrower focus, examining our own culture. More recently, the approach has been used with an even narrower focus: to learn what happens in an organization and understand the particular aspects of what seems familiar.

With these changes in focus, ethnography became more suitable to professional practices, such as design. In the US, the relationship between design and ethnography began in the 1970s, with the immersion of anthropologists in technology fields to examine how people worked and used tools in their workplaces[4]. This approach helped shed light on customer experiences and elicit deeper contextual information about people and their practices than conventional market research methods used at that time. This also opened the door for people to have more influential roles in the design process, generating the beginning of a *user-centred approach*. Ethnographers and designers continued working separately until, progressively, in the 1980s and early 1990s, a growing number of ethnographers started working in design environments and studying how products were used in their sociocultural context.

In the 1990s, *applied ethnography* emerged as an approach to specific problems involving practical application of findings[5]. In design, those working with applied ethnography didn't just focus on describing or analysing the context; rather, they sought understanding of the dynamic forces acting in those contexts and acted on that new knowledge[6]. By the mid-1990s, the use of applied ethnography in design was popularized and, although research studies were commissioned to external companies, many design firms added this service to their toolbox[7].

Towards the end of this decade, through a "quick and dirty" use of ethnographic methods, mostly for inspiration during early steps in the design process, the IDEO Company was instrumental in spreading the value of ethnography to other design fields. Since then, the creation, in 2005 by Ken Anderson and Tracey Lovejoy, of the annual Ethnographic Praxis in Industry (EPIC) international conference[8] has been an important milestone, allowing design ethnography practitioners from different fields to share their experiences.

More recently, a third transformation originated online—*virtual ethnography*—as the focus moved from looking at the "real" world to also studying the virtual world[9] (Chapter 5). This change has opened the applicability of the approach to studying other ways people interact through channels beyond face-to-face communication. In the virtual context, the meaning of "participate" or being in the "field" has changed and adapted to emerging

---

[4] E-Lab LLC (now part of Sapiens Corporation) and Xerox Park, both in the US, were two of the most influential organizations to use this approach.
[5] Anderson K., Salvador, T., & Barnett, B. (2013) *Models in Motion: Ethnography Moves from Complicatedness to Complex Systems*. EPIC 2013 Proceedings, 1, 232–249.
[6] Anderson, K. (March, 2009) Ethnographic research: A key to strategy, *Harvard Business Review*, 87(3), 24.
[7] Koskinen, I., Zimmerman, J., Binder, T., Redström, J., & Wensveen, S. (2011) *Design Research Through Practice: From the Lab, Field, and Showroom*. Waltham, MA: Morgan Kaufmann.
[8] Supported by Intel and Microsoft, the first EPIC Conference was held in 2005, [online], Available at: https://www.epicpeople.org/ [Accessed 26 December 2017].
[9] Flick, U. (2009) *An Introduction to Qualitative Research*, 4th edn, Thousand Oaks, CA: SAGE.

technology and virtual spaces accessible through the World Wide Web and electronic tools. The Internet has become a tool to study people you could not otherwise reach.

Perhaps because of the type of projects that information designers were working on until a few years ago, ethnography didn't reach the same level of popularity in the community as it did with product or engineering design. Information designers working in studios or companies focused on wayfinding projects are probably those most familiar with this research approach.

Contextual inquiry

Context inquiry emerged in the late 1980s as a data-gathering approach in design[10]. This technique combines field observations and in-depth interviews to gather a more holistic understanding of who people are and what they do. Contextual inquiry is rooted in the principle that people have difficulty articulating what they do, how they do it, and why they do it[11] (Chapter 4). This approach is often used in human computer interaction (HCI) to study the use of technology in its context of use and inform the software development process[12].

In the design research space, when using this approach, designers visit participants' workplace to observe and conduct one-on-one interviews. The goal of these interviews is to understand the actions performed by the participants and their motivations and strategies. Rather than being a classical question-answer interview, the interaction between designer and participant reflects a discussion or conversation, aimed at developing a shared interpretation of the work.

In information design, flavours of this approach are often used to gain an understanding of information-intense situations (e.g. understand how a help desk works or how people complete forms)[13]. This approach is also used to identify concrete details of how people use current designs and discover their problems, explore opportunities for people to work in new ways, or understand workflow.

---

10 Whiteside, J., Bennett, J., & Holtzblatt, K. (1988) Usability engineering: Our experience and evolution, in Helander, M. (ed.) *Handbook of Human Computer Interaction*, New York: North Holland.

11 Beyer, H. & Holtzblatt, K. (1998) *Contextual Design: Defining Customer-Centered Systems*, San Francisco: Elsevier, and Holtzblatt, K. & Beyer, H. (2014), Contextual Design: Evolved, *Synthesis Lectures on Human-Centered Informatics*, 7(4), 1–91.

12 Beyer, H. & Holtzblatt, K. (1998) *Contextual Design: Defining Customer-Centered Systems*, San Francisco: Elsevier; Madrigal, D. (2009) *Contextual Interviews and Ethnography: Two Different Types of Home Visits* [online], Available at: http://usabilitypost.com/2009/09/09/contextual-interviews-and-ethnography/ [Accessed 14 November 2017].

13 Beabes, M. & Flanders, A. (1995) Experiences with using contextual inquiry to design information, *Technical Communication*, third quarter, 409–420.

Usability testing

In the 1980s, usability testing, a research approach which builds on ergonomics and cognitive psychology, emerged from industrial and interaction design communities as a way to better understand the interaction between people and technology. The goal is to watch intended users using a specific design to complete tasks to identify usability issues. The work of Don Norman and Jacob Nielsen[14] was seminal and helped popularize usability methods.

In some cases, usability testing is conducted in the field to study the use of existing technology in context. However, differently from the first two approaches, usability testing is mostly conducted in a laboratory setting because it is easier for design researchers to control the variables under study. Consequently, these usability studies don't provide insights about the "context" or what users feel when using a given technology. Usability testing doesn't focus on examining the environment and the overarching experience in which a solution is meant to do its work.

Usability testing is quite popular among information designers who work on interactive projects such as the design of smartphone applications or other digital tools[15]. Recently, information designers have sought a more empathic research approach to determine the effectiveness and usability of a product based on users' emotions and experiences. This quest led to *user experience (UX) design*: a more expansive approach to measure usability, encompassing all aspects of a user's interaction with a system.

Participatory action research

In participatory action research all stakeholders (designers, researchers, developers, clients, intended audience, and other business decision-makers) actively work together to better understand a situation and generate solutions. Although roles blur and participants become a critical component of the process, they don't make final decisions nor work with the tools that designers or researchers use. In the design context, this approach emerged from Scandinavian countries in the 1990s as *participatory design* (today referred to as co-design)[16] for technology and software development, but Liz

---

14 Norman, D. (1988) *The Psychology of Everyday Things*, New York: Basic Books Inc.; and Nielsen, J. (1999) *Designing Web Usability: The Practice of Simplicity*, Indianapolis: New Riders Publishing. In 1998, Norman and Nielsen created Nielsen Norman Group: Evidence-Based User Experience Research, Training, and Consulting. [https://www.nngroup.com/ Accessed 10 November 2017].

15 Sless, D. (2012) Design or "design" – Envisioning a future design education, *Visible Language*, 46(1/2), 54.

16 Visser, F.S., Stappers, P.J., Van der Lugt, R., & Sanders, E.B.N. (2005) Contextmapping: Experiences from practice, *CoDesign: International Journal of CoCreation in Design and the Arts*, 1(2), 119–149.

Sanders brought this approach closer to the work of designers[17]. Utilizing this approach, designers organize workshops and sessions involving simple exercises and informal interviews in the form of conversations and discussions. Participants are seen as "partners"[18], not just users. They take an active role, working together with various techniques (e.g. prototypes or mock-ups, games, mood boards, storyboards, scenarios, personas) to provide input to the design process. Designers use this input to inform decision-making.

In information design, participatory design is often used in workshops to help find concepts or ideas for unframed projects, or identify parts of an existing design, that aren't working as intended.

Self-documentation

Originating from psychology and anthropology, *self-documentation* is a less intrusive approach than that of more traditional methods such as observations or interviews. This approach involves participants self-documenting their experiences and reflecting on their everyday activities, in their own time and environment without the researcher's presence. Participants use a diary or other materials to record entries or capture their thoughts and experiences.

Since the 90s, *Diary studies* have been increasingly used in design fields[19], such as HCI, to understand the use of new technology in the workplace. Participants respond to given questions by writing answers and logging in a diary when and how they interact with a given phenomenon or perform specific activities. In 1999, Bill Gaver, Tony Dunne, and Elena Pacenti[20] created *Cultural probes*, a qualitative design-led self-documentation method to gain deep understanding of people in different cultures, inaccessible to researchers. In contrast to diaries, cultural probes aimed at capturing general attitudes and behaviours, and better understanding participants' lifestyle rather than only daily interactions with a specific design. They also have a strong design component, needed to develop packages or kits with materials, and various evocative tasks to help participants self-report data.

In the early 2000s, Finnish designer and researcher Tuuli Mattelmäki broadened the use of probes, suggesting their use in the context of design practice to help designers build empathy with intended audiences. She combined the use of probes with interviews and the creation of artefacts to also involve the intended audiences in the process and gain a more holistic un-

---

17 Sanders, E.B.N. & Stappers, P.J. (2008) Co-creation and the new landscapes of design, *International Journal of CoCreation in Design and the Arts*, 4(1), 5–18.
18 Sanders, E.B.N. (2016) Design research in 2006, *Design Research Quarterly*, V. I(1), September, 1–9.
19 Bolger, N., Davis, A., & Rafaeli, E. (2003) Diary methods: Capturing life as it is lived, *Annual Review of Psychology*, 54(1), 579–616.
20 Gaver, B., Dunne, T., & Pacenti, E. (1999) Design: Cultural probes, *Interactions*, 6(1), 21–29.

derstanding. These *Design probes*[21] also involve a more structured means of gathering data that allows it to be analysed and used for both inspiration and information, helping designers make decisions in various stages of the design process.

Design probes and diary studies are sometimes used in information design in large or unframed projects involving the design of experiences or services. These methods are more commonly used, however, in UX design and service design fields. During the last years of the 20th century and beginning of the 21st, these human-centred research approaches have provided information design practice with a *qualitative* lens and a rich repertoire of methods to gain better understanding of people at various steps in the design process.

Understanding qualitative research

Qualitative research is complex and broad. Its goal is to understand the meanings of human behaviours and attitudes, based on subjective experience, in order to develop empathic understanding. Unlike *quantitative* approaches, qualitative research involves the collection and analysis of non-numerical data in the form of rich descriptions and explanations of how people understand the world. Qualitative studies seek answers to "what", "how", and "why" questions, by focusing on words denoting emotions and feelings, actions, rituals, experiences, perspectives, impressions, and opinions, and moving beyond the obvious to understand underlying needs and causes.

Although common types of research studies conducted in information design fall into the category of quantitative research and aim at testing a hypothesis (e.g. does the new design help users book a flight?), determining objective measures (e.g. what criteria can be established to assess performance?), or gathering large quantities of data to generalize results (e.g. how many target users find a design useful?), today there is a growing number of companies that offer some mode of qualitative research as part of their skill set. For example, user testing studies are often conducted for information visualization and interface design projects to determine usability and content issues with app designs[22]. In other information design fields, such as document design, empirical research studies are frequently conducted to learn how readers comprehend and use information[23].

21 Mattelmäki, T. & Battarbee, K. (2002) Empathy probes, in *PDC 02 Proceedings of the Participatory Design Conference*, pp. 266–271; Mattelmäki, T. (2008). *Design Probes*, 2nd edn, Vaajakoski: University of Art and Design Helsinki.
22 Marcus, A. & Jean, J. (2009) Going green at home: The green machine, *Information Design Journal*, 17(3), 235–243.
23 Schriver, K. A. (1997) *Dynamics in Document Design: Creating Text for Readers*, New York: Wiley.

Typically, when qualitative research is used in information design, its most commonly used form is open-questions embedded in a quantitative framework such as a questionnaire. These are *Small q studies*[24]. Despite their qualitative characteristic, these studies don't seek to understand how people construct meaning, because their starting points are pre-determined categories. In other cases, qualitative research is conducted under experimental conditions where participants go to information designers' studios, for example, to use and test a new website.

This book doesn't focus on these types of qualitative studies because they don't provide a deeper understanding of people's lifestyle and everyday experiences. Rather, the premise of the book is that information designers should adapt and use methods from the above qualitative approaches, and go to the field to collect data in participants' environments. For example, while usability testing is mostly conducted in designers' own studio, the approach can be adapted to be used in the field by the designer going to where the participant lives or works and observing how they use a new website. Going into the field will help information designers understand *why* something didn't work as planned or *how* people behave when performing a specific task.

Field studies, the types of qualitative studies needed in information design are *Big Q studies* (Figure 2.2). These begin with a hunch, open-questions, or just topics to be explored, and involve 'open-ended inductive research methodologies'[25] such as field research methods with a human-centred approach. These types of qualitative methods have some explicit structure, such as guiding questions, but aren't completely structured like a survey[26] (Figure 2.3). The focus is on capturing variations, documenting inconsistencies, and exploring individual differences between participants' experiences and outcomes. The goal is to capture whatever occurs, is seen and heard, including unexpected reactions or variations.

Why clients (and designers) don't trust qualitative research

Despite the benefits of research to enhance the quality of information design work, its use is still often questioned among practitioners and, particularly, the use of qualitative, contextual studies, like field research, are rare. As an example, many information design companies don't use qualitative

---

24 Willig, C. (2013) *Introducing Qualitative Research in Psychology*, London: McGraw-Hill Education.
25 ibid.; Braun, V. & Clarke, V. (2013) *Successful Qualitative Research: A Practical Guide for Beginners*, Thousand Oaks, CA: SAGE.
26 Blandford, A.E. (2013) *Semi-structured Qualitative Studies*, Interaction Design Foundation.

TYPES OF INFORMATION DESIGN RESEARCH

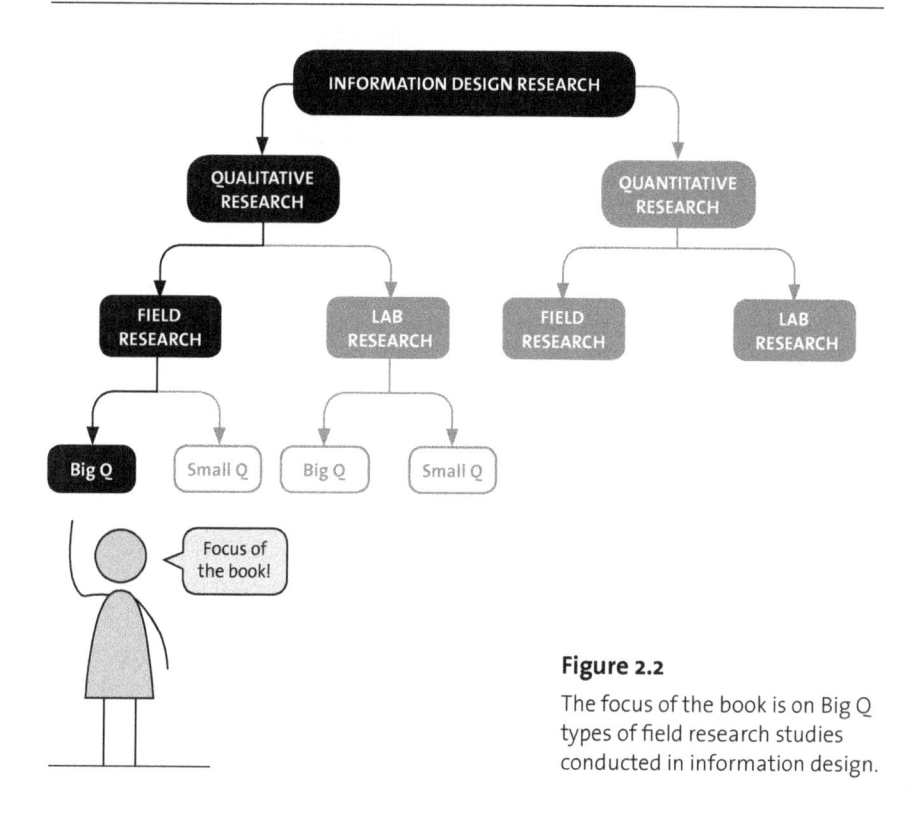

Figure 2.2
The focus of the book is on Big Q types of field research studies conducted in information design.

STRUCTURE IN QUALITATIVE METHODS

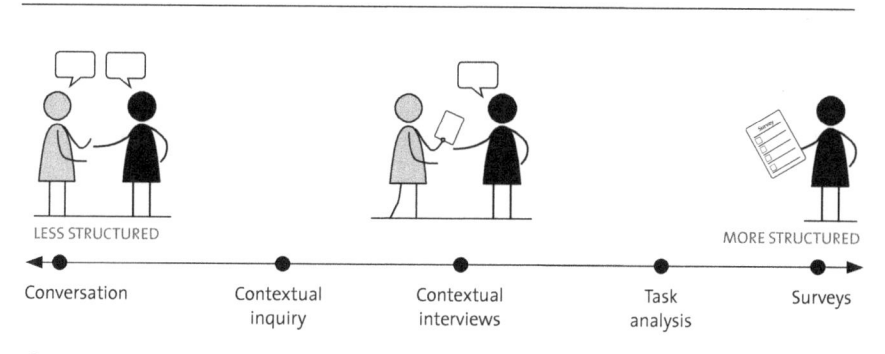

LESS STRUCTURED → MORE STRUCTURED

Conversation — Contextual inquiry — Contextual interviews — Task analysis — Surveys

Figure 2.3
Qualitative methods used in Big Q studies are semi-structured as you will have guiding questions that provide direction but these questions can change from one session to the other.

Bridging information design and field research | 25

research because they don't believe in it[27]. In other cases, those information designers using some form of field research as part of their regular practice need to "make the case" for it and convince clients of its value. Other information designers are sceptical of research in general and don't appreciate its value because they see it as only necessary to compensate for lack of experience. These information designers rely heavily on their years of design experience in the field and find research time-consuming. For them, any type of research also restricts their creativity. The following are common criticisms about the use of qualitative research in professional contexts:

- **Qualitative inquiry isn't useful**[28]: "Findings are not generalizable". Qualitative research is often described as too subjective, with no analytic depth, whose findings don't represent credible results because there are neither numbers nor an appropriate set of criteria to determine quality. This view corresponds to that of someone used to quantitative research. Quantitative findings are based on large samples, often used as "measurements" and "data" (numbers) to provide "objective" insights. This is a very different perspective from that of qualitative research.

- **Anyone can do it:** "Asking questions is easy". It is believed that everyone can research and collect data by asking questions and observing people doing something, while they just take some notes. This thinking challenges researchers' expertise and disqualifies the deep insight they bring.

- **Qualitative data is difficult to analyse:** "Yes, we use qualitative methods but analysing data is time-consuming and findings are hard to apply". Even among those who do appreciate the value of qualitative research and believe that it decreases the unknowns associated with redesigning or creating a new design, the tendency to cut corners on analysis, due to either financial- or time constraints, persists. The reality is that well-grounded interpretations and thorough analysis take time. When this part of the research process is rushed, resulting designs are ambiguous, unclear, and hard to understand or use. Even the richest of datasets would make no real difference to a design's quality, if it weren't properly analysed.

These views suggest that poor understanding of both what qualitative research entails and its benefits seems to be at the root of the problem. Qualitative research isn't common knowledge beyond academic contexts; it is mostly foreign to design professionals, corporates, and organizations. These

---

27 Ladner, S. (2014) *Practical Ethnography: A Guide to Doing Ethnography in the Private Sector*, Walnut Creek, CA: Left Coast Press.
28 Anderson, K. (March, 2009) Ethnographic research: A key to strategy, *Harvard Business Review*, 87(3), 24.

misconceptions may seem inconsequential acts within the bigger picture because many information design companies do research, but they have a cumulative effect with the potential to devalue qualitative work. If these misconceptions and tendencies perpetuate, they could lead to the wrong perception that qualitative research doesn't actually have much to offer after all[29].

In addition, *clients* play a key role in the infrequent use of qualitative approaches in information design. Some have a persistent view that quantitative research, such as market research, is all they need and struggle to understand how a more interpretative type of research can actually make a difference to the final outcome. Those who seem to understand their value are reluctant to allocate the necessary time and budget, because they are perpetually in a hurry and pressed for immediate results.

Regardless of whether they are information designers or clients, what stands between those who don't see the value in qualitative research and those who do is their **different points of view about reality**. When we encounter something that seems to present a view contradictory to our belief, our first reaction is to describe it as incorrect or invalid. Each research approach is rooted in what are called *paradigms, philosophical traditions,* or *points of view* about reality[30]. Each paradigm represents a way of reasoning and thinking about the world, making sense of its complexities; that is, people who have different points of view believe in different types of 'truths'[31]. When a point of view doesn't resonate with ours, we don't understand it. Paradigm-related biases are the source of misunderstandings and the reason why someone may dismiss the value of one research approach and not another.

There are many quantitative and qualitative research paradigms[32], each proposing a different explanation about reality that shapes the way we see and understand things. Particularly, this section discusses the views of *positivists* and *constructive-interpretatives* (Figure 2.4).

Many information designers, clients, marketers, product managers, and consultants are *positivists*: they judge what they consider "good" research and credible and valuable findings based on "facts" that they can observe—for example: how fast did people complete the new form? How many people downloaded the new app? How much faster is booking a flight when using the new interface? For many organizations, truth is provided by numerical data gathered via scientific methods, not by thick descriptions. The value from these descriptions is mostly unknown, undervalued, or unappreciated.

29 ibid.
30 Babbie, E. (2010) *The Basics of Social Research*, 5th edn, Wadsworth Publishing.
31 Ladner, S. (2014) *Practical Ethnography: A Guide to Doing Ethnography in the Private Sector*, Walnut Creek, CA: Left Coast Press.
32 Patton, M.Q. (2002) *Qualitative Research & Evaluation Methods*, 3rd edn, Thousand Oaks, CA: SAGE.

Figure 2.4

Common views of positivists and constructive-interpretatives.

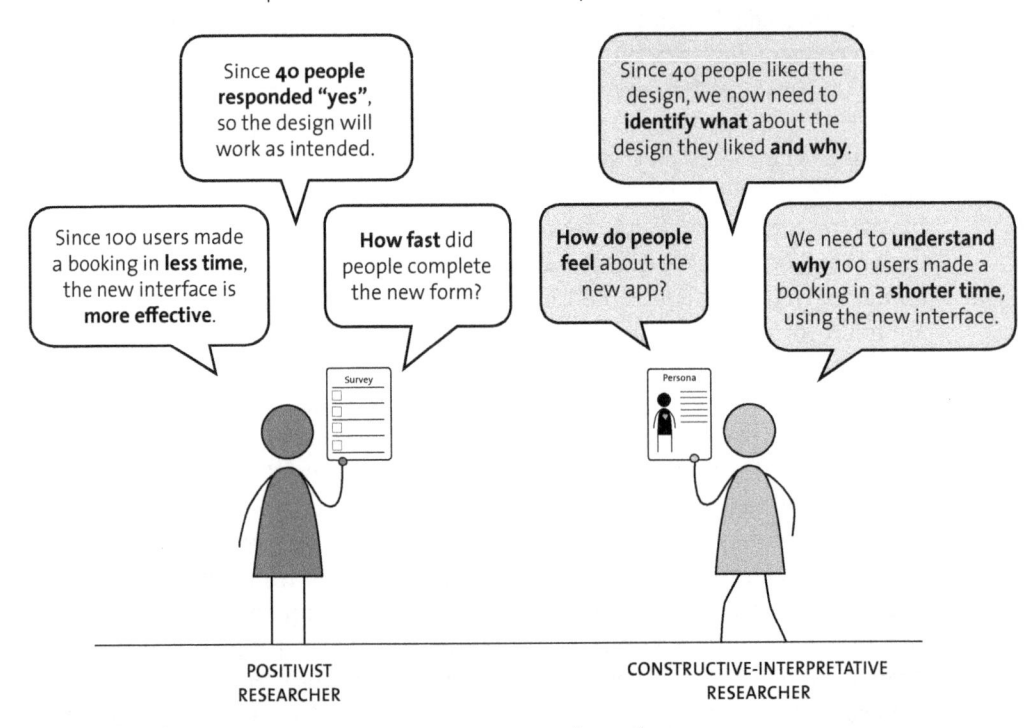

POSITIVIST RESEARCHER

Speech bubbles:
- Since **100 users made a booking in less time**, the new interface is **more effective**.
- Since **40 people responded "yes"**, so the design will work as intended.
- **How fast** did people complete the new form?

Assumptions:
- There is an objective reality or truth "out there" about a given topic.
- My role is to discover and present that reality or truth in an unbiased way.
- Dimensions and variables to measure should be agreed on before a study begins.

CONSTRUCTIVE-INTERPRETATIVE RESEARCHER

Speech bubbles:
- **How do people feel** about the new app?
- Since 40 people liked the design, we now need to **identify what** about the design they liked **and why**.
- We need to **understand why** 100 users made a booking in a **shorter time**, using the new interface.

Assumptions:
- There is a subjective reality that is shaped by my interpretations and those of the study participants.
- The interpretation process is how we make sense of reality.
- Dimensions and variables will emerge throughout the study as I interpret the data.

Positivists also follow assumptions that people's actions can be predicted, and that there is a right answer that leads to accurate predictions. In information design, these assumptions strongly manifest themselves during the evaluation of solutions. Some information designers and clients believe that the only way to rigorously measure whether an information design solution is successful after implementation is through numbers, such as data collected with Google Analytics or surveys. While numbers and statistics are important because they provide hard evidence of overall effectiveness, numbers alone don't suggest the reasons why a design works or doesn't work, nor show human experiences or specify whether the context of use affects the performance of a solution in any way.

In contrast, *constructive-interpretatives* follow a different point of view from that of positivists: they believe that, while you can make inferences and educated guesses based on experiences and patterns, people's dynamics and interactions are too complex to make radical judgements such as right or wrong responses. Instead of looking for "observable events", this paradigm looks at what something means to people—for example: what was the experience of completing the new form for each person? How do people feel about the new app? To answer these questions, rather than seeking facts or right answers, constructive-interpretative researchers look for descriptions of human experience and meanings manifested through stories, behaviours, feelings, and opinions. In other words, they seek depth of understanding.

People find meaning in every situation, for example when navigating a space or understanding how to use a device or interpreting an infographic. The overall experience involves both how it makes them feel ("This is nice, I like it") and what they can learn from it; that is, how they extract understanding from it ("To start using the remote control first I need to press the red button"). In most cases, this matters most to them, regardless of whether they find an exit quicker or achieve their goals in a shorter time. To identify the meanings that people attribute to information design solutions, information designers need more than just a quantitative approach; they need one that helps understand people holistically.

Circling back to the earlier discussion, Big Q studies, such as field studies, involve the use of qualitative methods within a *qualitative paradigm*, such as the constructive interpretative paradigm. These methods place people at the centre, to understand their social world. Insights gathered, such as people's experiences, satisfaction, and emotions, can add significantly to traditional questionnaires, creating more actionable recommendations, and helping information designers make more objective decisions and give credibility to their claims. This type of qualitative insights gives information designers a theoretical supportive framework—the *why*—that even experienced designers don't have.

Working with field research: the information design process revised

These are some expected questions that may emerge from the previous discussion: how can information designers work with this type of qualitative research? Where in the process would these methods fit? To answer these questions, an information design process is introduced, with a stronger *research emphasis* to reinforce specific steps and help information designers effectively deal with growing complexity and better understand people.

This process combines existing theoretical work[33,34], observations of design colleagues, learnings from many cohorts of students, and my own experience as a professional information designer and researcher.

This research-led information design process is composed of a series of steps and associated activities (e.g. describe, explain, instruct, combine) that an information designer performs to make sense of a problem, find a solution, make the solution a reality, and test its effectiveness. It also accounts for field research studies at various of those steps to inform decisions made in subsequent steps throughout the process. Some research studies focus on the human perspective to study people's behaviours, needs, and frustrations, independently of whether they are interacting with a solution, while others focus on the specific type of relationship between people and a solution, and on understanding that relationship[35]. This process complements more traditional design activities with *applied research*[36].

While not all information designers work in the exact same way, most perform similar steps or execute them in the same order. At a broad level, they go through two stages: **conceptual design**, the "figuring out" stage, involving exploration, analysis, and making sense of a problem in order to find a solution, and **prototype design**, the "making stage", centred on the detailed design, prototyping, implementation, and evaluation of the solution. Three modes of activities are present throughout each of these stages:

- *Exploring or diverging* to learn, gain understanding, and evaluate ideas.
- *Analysing or converging* to extract meaning, select, and make decisions.
- *Creating or synthesizing* to develop ideas into tangible outputs.

Figure 2.5 presents the suggested research-led information design process. Although the process is shown as in a linear way, it isn't strictly linear, as cycling back and repeating steps are necessary actions to gather data, add

---

33 There are two strong views about how information designers work: (1) those who describe the process as involving an 'Aha! moment', characterized by pure inspiration or intuition, for example: Klanten, R., Bourquin, N. & Ehmann, S. (2008) *Data Flow, Visualizing Information in Graphic Design*, Berlin: Gestalten, and (2) those who argue that information designers follow a process with identifiable actions and steps, for example: Jones, C.J. (1992) *Design Methods*, 2nd edn, New York: John Wiley; Cross, N., Christiaans, H., & Dorst, K. (1996) *Analysing Design Activity*, Chichester: John Wiley & Sons; Dorst, K. & Lawson, B. (2009) *Design Expertise*, Oxford: Architectural Press.

34 Frascara, J. (ed.) (2015) *Information Design as Principled Action: Making Information Accessible, Relevant, Understandable, and Usable*, Champaign, Il: Common Ground Publishing LLC.

35 Sless, D. (2008) Measuring information design, *Information Design Journal*, 16(3), 250–258.

36 Applied research differs from basic research, commonly used in academic contexts. The latter is initiated by scientific curiosity and focuses on expanding knowledge and discovering the unknown but isn't focused on the practical application of findings. The former focuses on solving practical problems of the modern world by investigating a specific problem situation (e.g. specific set of circumstances, specific users), and findings are directly applied in response to a specific situation such as the creation of a solution.

new insights, and refine a solution after testing and evaluation. The stages and steps of this process are unpacked below.

The beginning: conceptual design

The effectiveness and impact of any information design work depends on the orchestration of a number of factors, but it begins with a solid foundation. Particularly, information design places a strong emphasis on this initial stage, *conceptual design*, distinguishing this field from graphic design and other design fields. Conceptual design focuses on the definition and exploration of the problem and conception of ideas[37]. For that reason, information designers devote considerable effort to this stage, moving through **five core steps** and a series of activities to define and understand the initial problem before trying to solve it[38]. Throughout conceptual design, they cycle many times through the steps, starting with more general and moving to more detailed cycles[39]. A summary of each step follows:

1. **Problem understanding:** The first step focuses on identifying, understanding, and framing the problem by defining questions that should be addressed by a solution at the end of the process. In the last decade, growing attention has been paid to this step[40]. Information designers approach it from various directions; some follow an explicit and well-defined sequence of activities, while others adopt a more informal and less structured approach[41], but field research has become an essential component. This is even more necessary when dealing with unframed problems or fuzzy situations, to manage their ambiguous and chaotic nature. Chapter 5 provides methods to support this step.

2 & 3. **Subject matter & audience understanding:** The second and third steps tend to occur in parallel, as they focus, respectively, on gaining a deep understanding of the subject matter and the audience. To understand the relevant topic, information designers move through information cycles, combining desk research and their own experience with first-hand information from clients and audience. Field research is used to explore complex topics and identify audience needs. The more complex and specialized the challenge is, the greater the need to gather

---

37 Pontis, S. (2012) *Guidelines for conceptual design to assist diagram creators in information design practice*, PhD thesis, University of the Arts London, UK; Ware, C. (2008) *Visual Thinking for Design*, USA: Elsevier; Wurman, R.S. (1996) *Information Architects*, Zurich: Graphics Press.
38 Ware, C. (2008) *Visual Thinking for Design*, San Francisco: Elsevier.
39 Jones, C.J. (1992) *Design Methods*, 2nd edn, New York: John Wiley.
40 Sanders, E.B.N. & Stappers, P.J. (2008) Co-creation and the new landscapes of design, *International Journal of CoCreation in Design and the Arts*, 4(1), 5–18.
41 Ware, C. (2008) *Visual Thinking for Design*, San Francisco: Elsevier, 156–157.

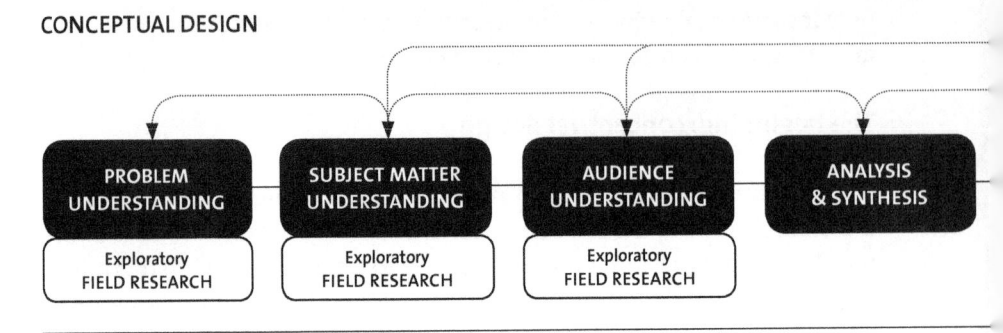

CONCEPTUAL DESIGN

```
┌─────────────────┐   ┌─────────────────┐   ┌─────────────────┐   ┌─────────────────┐
│    PROBLEM      │   │ SUBJECT MATTER  │   │    AUDIENCE     │   │    ANALYSIS     │
│  UNDERSTANDING  │   │  UNDERSTANDING  │   │  UNDERSTANDING  │   │   & SYNTHESIS   │
└─────────────────┘   └─────────────────┘   └─────────────────┘   └─────────────────┘
   ┌───────────┐         ┌───────────┐         ┌───────────┐
   │Exploratory│         │Exploratory│         │Exploratory│
   │  FIELD    │         │  FIELD    │         │  FIELD    │
   │ RESEARCH  │         │ RESEARCH  │         │ RESEARCH  │
   └───────────┘         └───────────┘         └───────────┘
```

EARLY STAGE: Empathy and Inspiration

RESEARCH GOALS
- Determine project focus or direction
- Gain familiarity with basic facts, key parties, and gaps
- Create a general picture of the problem
- Formulate questions for next steps
- Identify new areas for exploration and gathering more focused data

RESEARCH GOALS
- Expand initial understanding of a situation
- Gather detailed insights to create more accurate picture
- Identify key components (steps, stages, tasks, etc.)
- Clarify sequence or flow of steps in a process

Figure 2.5
Research-led information design process. The information design process can be supported at specific steps by exploratory or evaluative field research.

information from subject experts. Learnings are used to determine criteria for making decisions throughout the process. Chapter 5 provides methods to support these steps.

4. **Analysis & synthesis:** At the end of the first three steps, large amounts of content from different sources compose the raw datasets for use as the basis for creating a solution. The next goal is to determine what is to be designed and sometimes what should not be designed. During this step, raw datasets are analysed to identify and extract any specifics or nuances about the project at hand that can help address the initial questions. Guidance and methods provided in Chapters 7, 8, and 9 could be used to support this step.

5. **Concept design:** The final step involves the visualization of learnings synthesized from the previous steps, the generation of ideas to answer the initial questions, and the creation of concept solutions or proposals for some of those ideas. By the end of conceptual design, the format that the deliverable could take should be mostly defined: a product, a ser-

PROTOTYPE DESIGN

```
CONCEPT          DETAIL
DESIGN           DESIGN          IMPLEMENTATION     EVALUATION

Exploratory or Evaluative    Exploratory or Evaluative                    Evaluative
FIELD RESEARCH               FIELD RESEARCH                               FIELD RESEARCH
```

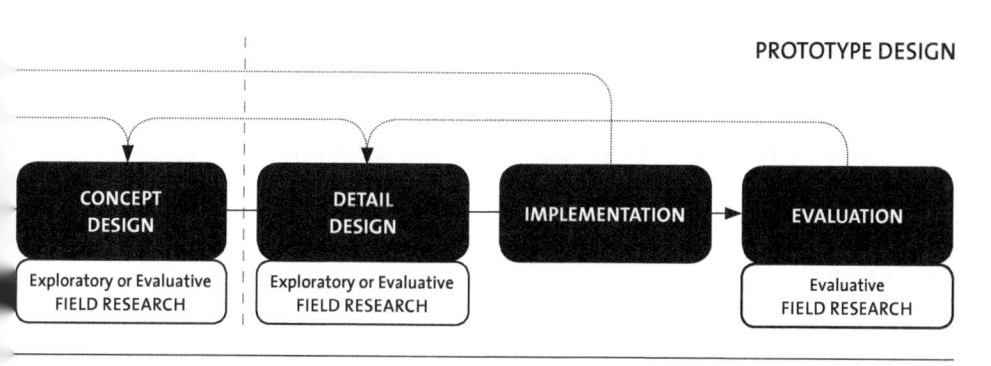

MIDDLE STAGE: Inspiration and Improvement				**LATE STAGE:** Improvement

RESEARCH GOALS

- Elaborate and enrich an idea or concept
- Validate an idea or concept
- Identify areas for improvement
- Test a concept or an unfinished prototype

RESEARCH GOALS

- Test a functional prototype
- Optimize a functional prototype
- Determine effectiveness of a design solution

---

vice, a system, a poster, a strategy, an interface, etc. *Formative field evaluations* to gauge audience's initial responses to a design concept have started to become commonplace in this step of the process to identify flaws in a design idea early on and provide sufficient time to start over if necessary or keep moving ahead more confidently. Chapter 6 provides guidance and tools for this step.

The execution: prototype design

During this stage of the process, ideas and proposals developed during conceptual design are executed at higher quality, produced, tested, and implemented. Information designers move through **three core steps** and a series of activities to execute their ideas. A summary of each step follows:

> **6. Detail design:** This step focuses on the development of a prototype based on a chosen design concept. Formal and visual possibilities are explored, including decisions about layout, text, typography, colour, visual language, and other design specifications that support the aims of the project and respond to research findings obtained during conceptual design steps. This step also involves the testing, through *formative field evaluations*, of the key aspects and appropriateness of not fully functioning prototypes. Before moving forward in the process, these evaluation

studies help determine whether the direction of a design is in line with the goal of the project and audience needs. Increasingly, these evaluations are being conducted in the field to gather richer and more accurate insights. These are then used to make major changes and generate a new prototype or make minor adjustments to optimize the design. Chapter 6 provides guidance and tools for this step.

7. **Implementation:** At this step of the process, the design solution has been fully produced and evaluated and is ready for use. Implementation involves launching or delivering the final design, for example by pushing an interactive data visualization live on a website, distributing a print instruction guide with the product it accompanies, or unveiling a new wayfinding system in a public space. There may also be supporting materials and activities to help familiarize the audience with the new solution, as with an updated website or redesigned utility bill. Often, information designers aren't directly involved in implementation; they may deliver the final product to the client or third party, who is then responsible for launching the solution to the audience.

8. **Evaluation:** After a design solution has been implemented for a set time, *summative evaluations* can be conducted to assess its overall performance. This is another step where the need for field research evaluations has grown, as more traditional evaluations like A/B testing don't work for information design solutions to unframed challenges. Chapter 6 provides guidance and tools for this step.

When to conduct field research in information design

You can conduct a field study for many reasons, but the following two will be the most common ones: *explore* to understand an unknown situation, describe further a known situation or explain why the situation occurs[42], or *evaluate* to learn how something is received or how something works. A field study often has more than one purpose (e.g. understand and describe), but one is dominant. These purposes align to specific steps of the information design process and indicate when field research is more helpful:

- **Define direction or frame the problem** (Process step 1). The focus of the study is *exploratory* either because the topic is very new, you know little or nothing about it, or the problem is unclear and undefined. This is often the case when work starts on an unframed challenge and you don't

---

[42] Lawrence, N.W. (2013) *Social Research Methods: Qualitative and Quantitative Approaches*, 7th edn, Harlow: Pearson New International Edition.

have a clear picture of the situation or where to start. Your goal with this type of study is to gather enough information to formulate more precise questions to use to move forward in the information design process, to either frame the problem or design another more focused and extensive study (e.g. understand your audience). Exploratory studies can be hard to conduct because they can be vague, with few guidelines and undefined steps. Exploring various information sources (oral, written, and visual) helps define initial directions to investigate. For examples of how this type of study was used in information design projects see Case Studies 2 and 4 in PART IV.

- **Understand people and identify needs** (Process steps 2 & 3). The study's focus is *exploratory* but the goal will depend on whether you already have a well-defined question, situation, or direction to investigate or you don't know the root of a problem[43]. Although you may have an initial understanding, your goal is to dig deeper, to put together a more accurate picture of your potential audience with specific details about their behaviours, a situation, or to identify causes of something occurring. Insights will give you more confidence, inspiration, and a clearer direction to start generating ideas because you will have evidence to support design decisions. For examples of how this type of study was used in information design projects see Case Studies 1, 2, 3, 4, and 5 in PART IV.

- **Evolve a design idea, or validate an idea or unfinished prototype** (Process steps 5 & 6). The focus of these studies can be either *exploratory* or *evaluative* depending on whether you are trying to find inspiration to evolve an already effective idea or trying to determine whether an idea will be effective. Conduct exploratory studies when you already have a design concept but would like to find other or more ways to make the idea in line with audience's needs. Conduct the study with evaluative focus when you have already generated design concepts but are unsure whether they are on the right track or appropriate for the intended audience. For examples of how these types of studies were used in information design projects see Case Studies 2, 3, 4, and 5 in PART IV.

- **Optimize a finished design** (Process step 8). The study's focus is often *evaluative*, as you are trying to test the effectiveness of a final solution by determining whether it is accomplishing what it was designed to do. Conduct this type of evaluation study when you already have a working prototype of the idea and want to assess whether it helps the intended audience accomplish their aims. For examples of how this type of study

[43] ibid.

was used in information design projects see Case Studies 2, 4, and 5 in PART IV.

For some projects, field research alone isn't enough because the situation is too undefined, particularly, making it difficult to conduct this type of research at the beginning of a project. In such cases, or to gather a more complete understanding of a situation, an option is to combine field research with a form of quantitative research or a Small q method, such as a survey, and conduct a *mixed study*. As a result, you gather both quantitative and qualitative insights about the same situation.

Before you start to plan a field study for your information design project, the next chapter provides a deeper understanding of what this type of research involves, what components are necessary to add rigour to a method and credibility to a study, and what other key considerations you need to think about.

3 What is field research?

Field research encompasses many specific techniques and methods, but all closely examine people's environments. This means that these studies are conducted in *natural settings*: as the researcher—you! —will go into the field to where participants live or use a design, such as their homes or workplaces. Fieldwork gives direct and personal contact with participants and the situations studied. The concept of "going into the field" has expanded, with the Internet and social media used as research tools that help ask questions and observe people without physically being in the same place (Chapter 5).

Understanding key components

Empathy, interpretation, and ambiguity are key components of field research.

Empathy

Your goal with field research is to feel what your participants feel. This is building empathy. To achieve this goal, you immerse yourself in other's people's lives and in the types of situations they encounter so you can experience similar emotions. The adoption of this empathic approach is necessary to reduce any form of bias and judgement towards their lifestyles and viewpoints. It also generates closeness with people and helps better understand their feelings. To build empathy, you must be truly curious when you are in the field asking interview questions, listen attentively to what participants say, and observe each situation as if it is the first time you've seen it. If you can develop empathy for participants, by the end of a study, you will have an empirical understanding of their feelings. This knowledge puts you in a better position to create information design solutions that address their needs.

Interpretation

In field research, you use *inductive analysis*[1] to make sense of what was gathered and seen. This type of analysis involves interpretation of what partic-

---

[1] In contrast, quantitative research uses *deductive analysis*, as data is collected to confirm or refute an initial hypothesis or pre-defined theory (top-down approach). Data is analysed according to an existing or pre-determined framework (e.g. specific categories).

ipants said, did, and made[2]; it moves from specific observations to broader generalizations, theories, or conclusions (bottom-up approach, so up from the data). The focus of inductive analysis is on discovering patterns, themes, and categories in participants' data, rather than determining these prior to collecting and analysing data.

Ambiguity

Another intrinsic component is ambiguity because many variables aren't under your control. To some extent, you depend on participants' actions and behaviours. Tolerating uncertainty and accepting ambiguity are also significant aspects of this type of qualitative research, particularly important when analysing data, in helping resist the temptation to rush to conclusions. When making sense of quantitative data, the process can be more straightforward, as you can rely on percentages and statistics to construct understanding and identify trends, but when making sense of qualitative data, the process can be more uncertain. Later in this chapter, there is a discussion of criteria to increase confidence in a study and identified findings, minimizing uncertainty. However, as interpretation is the core of field research, there will always be a degree of ambiguity.

Considerations for conducting field research in information design

Going into the field to learn about people and their lives requires preparation and understanding key aspects that may influence the experience. The following are five of them, summarized in Figure 3.1.

Understand your role

In field studies, you have a fundamental role in the process because you are the principal instrument of data gathering. Before, during, and after the study, you need to pay attention to your own personal biases and assumptions towards the participants, the situation, and the context in which you conduct the study. This awareness of any possible effects that you may have on the research is described as *reflexivity*[3]. You should constantly assess your

---

2 Patton, M.Q. (2002) *Qualitative Research & Evaluation Methods*, 3rd edn, Thousand Oaks, CA: SAGE.
3 Patton, M.Q. (2002) *Qualitative Research & Evaluation Methods*, 3rd edn, Thousand Oaks, CA: SAGE; Blandford, A., Furniss, D., & Makri, S. (2016) Qualitative HCI research: Going behind the scenes, *Synthesis Lectures on Human-Centered Informatics*, 9(1), 1–115.

Figure 3.1

Five key considerations for conducting field research in information design.

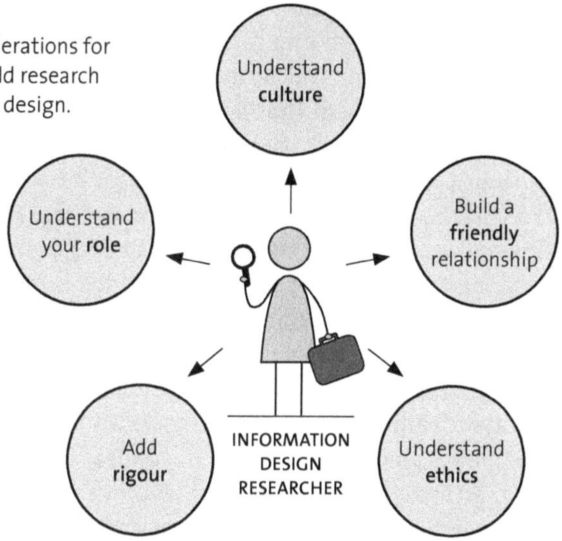

own contributions to the study and possible influence on participants or findings. One way of doing this is keeping a personal journal, index cards, or a Word document where you write any thoughts and feelings that emerge during the study. Another way is writing all your initial assumptions before going into the field. This will help you be more aware of them and minimize their influence in the study.

Understand culture

There is more to field research than simply following a research strategy and choosing a given method to gather and analyse data. As discussed in Chapter 2, behind any research decision, there is a paradigm which influences and gives direction to the whole research process. The *cultural context* of a study can also significantly impact its findings. It is essential to understand the role that *culture* plays in people's experiences, as it influences what we do and say, and how we think[4].

Cultural knowledge. Field research is rooted in context and, therefore, in culture. Implicitly or explicitly, culture impacts the events, motivations, behaviours, and attitudes that you study. When people make decisions, talk, and behave, they display their cultural background and also the specific social context in which they are immersed. For example, people behave dif-

---

[4] Ladner, S. (2014) *Practical Ethnography: A Guide to Doing Ethnography in the Private Sector*, Walnut Creek, CA: Left Coast Press.

ferently at home with family members than in the office with co-workers. To access people's inner world, you must observe their external behaviours (what they say, do, create), draw conclusions on their explicit, external behaviour, and infer the meaning of that behaviour. This means identifying the thoughts or feelings that may have triggered certain behaviour or word choice. Simultaneously, these inferences are influenced by your own *cultural knowledge* and biases. We acquire our cultural knowledge by interacting with our family and members of the community, listening to the radio, going to school, or observing people. Some of this knowledge is explicit (you are fully aware of it) but other parts are tacit (you don't really know why you do some things) [Chapter 4 expands this discussion]. Throughout our lives, cultural knowledge manifests itself in songs, sayings, idiomatic expressions, stylistic decisions, ways of behaving, tastes, preferences, and goals[5]. During fieldwork, you should pay attention to these same cultural dimensions in your participants and then analyse them to learn what they need. Another way to identify how culture influences people's decisions and interactions is to specifically look at their values, beliefs, and behaviours. A successful information design solution is in line with your intended audience's cultural knowledge.

Methods and culture. These relate to reflexivity and your biases. If planning to conduct a study in an unfamiliar culture, before making any research decisions, you should become familiar with that culture. This will help you determine what methods would be more appropriate. Firstly, communication may be an issue, unless you are proficient in that language. Secondly, participants from different cultures respond differently to the same methods and techniques; not taking this into account and making any necessary adjustments beforehand may give you a few headaches while conducting the studies. For example, I experienced this situation when trying to conduct a research study in Spain, following pretty much the same steps that I had successfully used a month earlier for a group of people in Finland. When tasks and activities didn't work out as originally planned, I realized that these two very different cultures were responding to the same prompts in quite distinct ways; particularly, I noticed significant changes in time frames for completing tasks, types of response, and level of engagement. Also, extremely introverted cultures may not respond well to face-to-face interviews as they may feel invaded. In this case, a more unobtrusive method (such as a self-documentation study) would be more appropriate, followed by email interviews.

Furthermore, it may be extremely hard to interpret intentions, expressions, gestures, or moods from a culture that isn't yours or to ask the right questions. While some gestures and expressions may be common to many

---

5 ibid.

countries, each region and country (and even city) has its own regional gestures. For example, in the UK the nose tapping gesture indicates conspiracy: 'Keep it dark, don't spread it around', but this meaning changes in Italy, where it just signals 'a friendly warning: "Take care, there is danger"'[6]. You may consider partnering with someone of the culture you are studying or choosing a participatory action research approach to invite people from the foreign culture into the design process.

Another aspect to consider is whether to recruit cross-cultural participants for a study. For example, this would be the case if working on a project involving a new design for a large city with a multicultural population as described in Case Study 2 in PART IV. In this case, the appropriate methods would depend on various factors such as the purpose of the study, whether you know where participants will be coming from, their level of understanding of the study, and willingness to participate. Also, be aware of the anxiety that participants from different cultures may experience when involved in a field study. For example, in culturally unknown scenarios, you may consider first using exploratory approaches to gain a holistic understanding of participants, rather than starting the study by testing a solution.

Build a friendly relationship

In most field studies, your presence in the setting can be intrusive or disruptive. To minimize disruption, pay particular attention to the relations between yourself, your team, and the participants. It is important to build *rapport* with your participants and the team early on[7]: develop a harmonious connection. This helps participants develop trust and feel comfortable in your presence. Ways of helping participants feel this way are learning their language, and organizing face-to-face meetings prior to the study to share experiences and jokes without judging them. In addition, this will help you start seeing and feeling their experiences from their perspective rather than yours, and breaking with your assumptions. Another way of building rapport is reassuring participants that you aren't assessing them at any point. Explain as many times as necessary that the study's goal is to learn about them or assess the functionality and effectiveness of a design.

In some cases, building rapport can take time and may not happen from one day to the next; some participants may need more time than others to open up or feel comfortable. Even in cases where you won't spend long periods with participants in their environments, as in self-documentation

---

6 Morris, D. (1979) *Manwatching: A Field Guide to Human Behavior*, 3rd edn, Frogmore: Triad/Panther Books.
7 Lawrence, N.W. (2013) *Social Research Methods: Qualitative and Quantitative Approaches*, 7th edn, Harlow: Pearson New International Edition.

studies, aim to develop a friendly relationship with each of them and be genuinely interested in what they share with you.

Understand ethics

As in any study involving people, the first rule is to avoid doing any physical or psychological harm. The following ethical considerations are key to meeting this rule[8]:

- Inform participants of the purpose of the study and what is expected from them (e.g. time commitment, tasks).
- Explain the right to withdraw at any time without reason or consequences.
- Ensure the confidentiality and anonymity of any data collected (even if you won't be writing a report afterwards), and who will get access to the data during the study.
- Explain how collected data will be stored and used, and whether or when it will be destroyed.
- Explain how findings will be shared and reported.
- Explain to participants that they will be debriefed at the end of the study if this couldn't be done in full at the beginning (e.g. to avoid bias).

Share the above information with participants in an *informed consent form*. This form has two goals: inform participants about the study and their rights, and gain their consent to be part of the study. Obtain participants' written consent by email or a hard copy before the beginning of a study session. If a participant doesn't give you consent, or changes their mind after the session and doesn't what you to use their data, you are obligated to delete that dataset.

In addition to these ethical considerations, plan in advance for keeping data private and confidential, your safety, and what compensation you would give participants.

Privacy and confidentiality. Respect participants' privacy and confidentiality throughout each step of the study (data gathering, analysis, and reporting), even among your team and other stakeholders (e.g. clients). Anonymize data as soon as possible: as early as when you start note-taking in the field. Ways

---

8 Patton, M.Q. (2002) *Qualitative Research & Evaluation Methods*, 3rd edn, Thousand Oaks, CA: SAGE; Blandford, A., Furniss, D., & Makri, S. (2016) Qualitative HCI research: Going behind the scenes, *Synthesis Lectures on Human-Centered Informatics*, 9(1), 1–115.

to do this are replacing people's names with pseudonyms (e.g. using made-up names) or with numbers (e.g. "P3"), and removing any other information that could identify participants (e.g. company names, colleagues' names, specific places, project titles, etc.). To ensure privacy, some companies only keep participants' data for a specific amount of time, such as 90 days or until the project is completed. Equally important is to keep confidential the new design you are testing until it is ready, if this is your case. For that, you may consider also asking participants to sign a non-disclosure agreement (NDA) at the beginning of the study.

Your safety. Regardless of the length or location of the study, take any necessary precautions to avoid potential unpleasant situations. For example, meet participants for the first time in public spaces or with a team member. If you must go to workplaces or homes, work in pairs and always schedule meetings during working hours or daytime.

Compensation. As far as possible, allocate some budget to compensate participants and others involved in the study for their time. Compensation can take any form, from cash to gift certificates. But avoid giving participants too substantial an incentive, to ensure they're not just participating for the compensation.

Add rigour

The key to gaining the most value from a field research study is *rigour*, which can take different forms—it doesn't necessarily have to be the 'logical accuracy and exactitude' demanded by academic contexts[9]. Research in professional contexts follows what is called *studio rigour*: a term introduced by Wood[10] to reflect a shift from academia to a form of rigour more in line with design practice needs. Studio rigour manifests itself in information designers' commitment and ethics when performing the work, and in the use of methods in a disciplined, deliberate manner to yield the greatest benefit and utility. For example, observational studies should be conducted in a systematic way, documenting everything observed with sufficient detail that, later, you or another team member can extract rich meaning.

---

9 Russell, K. (2002) Why the culture of academic rigour matters to design research: Or putting your foot into the same mouth twice, *Working Papers in Art and Design*, 2 [online], Available at https://www.herts.ac.uk/__data/assets/pdf_file/0007/12310/WPIAAD_vol2_russell.pdf [Accessed 17 January 2018].

10 Wood, J. (2000) TThe culture of academic rigour: Does design research really need it?, *The Design Journal*, 3(1), 44–57(14).

How to develop field research sensibility

Just as you are unlikely to make it to the end of a marathon if you haven't trained for a while, it is unlikely that you'll be a confident researcher the first time you go into the field. Conducting an effective interview or meaningful observational study requires training and practice. Even for seasoned information designers, training, patience, and practice are a must to build field research sensibility.

One way to start developing the necessary skills is to learn from other more experienced researchers. For example, accompany another researcher on several field sessions as observer or notetaker. Start training your eye; for example, practise observations when commuting by train or walking down the street. Another way is asking more questions (e.g. to clients, colleagues, friends, family); this will help you become a better listener too.

How to ensure quality and validity in field research

One of the biggest differences between qualitative and quantitative research is that the latter contains well-established methods, formulas, and rules to make sense and analyse data. Many of these methods operate with numerical data, gathered through pre-defined initial categories or frameworks, making the analysis "objective" and "quantifiable". In contrast, qualitative analysis involves a high degree of interpretation and depends on artful pattern recognition. The quality and validity of findings depend on the quality of the entire process, from the selection of methods and design of the study to the process of data gathering, analysis, and reporting.

Criteria for assessing the quality and validity of a study vary, based on the research paradigm (Chapter 2). So, the same criteria used in quantitative research shouldn't be used to judge trustworthiness in a field study[11]; for example, absolute objectivity is impossible to achieve in qualitative research. Here, five criteria—credibility, transferability, dependability, confirmability, and applicability—relevant to field research are reviewed and strategies to address each provided. Each criterion is presented as a question that information designers frequently ask themselves when conducting research (Figure 3.2). Addressing each question should increase your confidence in what you find and the overall quality of the study.

---

[11] Lincoln, Y.S. & Guba, E.G. (1985) *Naturalistic Inquiry*, Newbury Park, CA: SAGE.

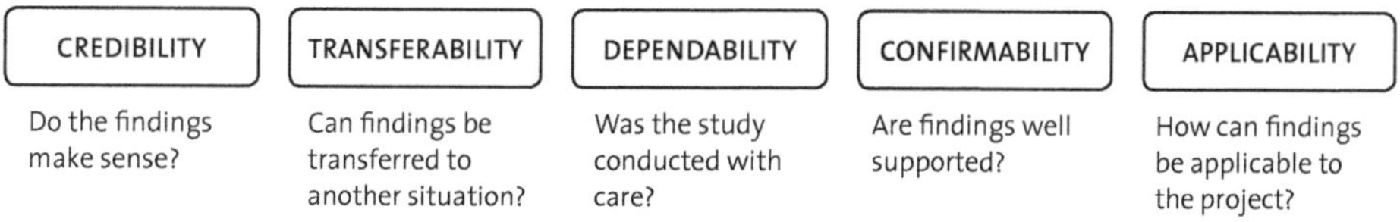

CREDIBILITY	TRANSFERABILITY	DEPENDABILITY	CONFIRMABILITY	APPLICABILITY
Do the findings make sense?	Can findings be transferred to another situation?	Was the study conducted with care?	Are findings well supported?	How can findings be applicable to the project?

Figure 3.2
Five quality and validity criteria for field research.

1. Do the findings make sense?

Credibility refers to the *internal validation* of the study and deals with whether findings are credible for stakeholders and clients, and provide an understanding of the situation under investigation from participants' viewpoints. Many strategies can be followed to promote confidence about findings. Some relevant for information design are:

- **Assess findings with participants.** Participants can help add credibility by assessing the accuracy of your interpretations and validating findings. As the study progresses, you can share emerging themes and partial conclusions with some participants randomly selected. For example, this is common in participatory action research, where participants' feedback is given throughout the process.

- **Use triangulation.** The term "triangulation"[12] refers to cross-verification of findings by approaching the same situation under investigation from more than one angle (more than two would be ideal! But two is better than one). The goal of triangulation is to identify and understand when and why differences occur. For example, obtaining different findings after using two distinct methods, such as observations and interviews, wouldn't invalidate any of them; it would just show different sides of the object under investigation. The key is to understand the reasons for those differences and construct a more complete picture. Table 3.1 describes three types of triangulation you can use to increase findings' credibility.

- **Document personal views.** As discussed earlier, keep a personal journal to acknowledge your assumptions during data collection and analysis.

[12] Patton, M.Q. (2002) *Qualitative Research & Evaluation Methods*, 3rd edn, Thousand Oaks, CA: SAGE; Blandford, A., Furniss, D., & Makri, S. (2016) Qualitative HCI research: Going behind the scenes, *Synthesis Lectures on Human-Centered Informatics*, 9(1), 1–115; Miles, M.B., Huberman, A.M., & Saldaña, J. (2013) *Qualitative Data Analysis*, SAGE.

TYPES OF TRIANGULATION

Methods triangulation	Gather data with multiple qualitative and quantitative methods, e.g. interviews, surveys, observations, and diary studies.
Data triangulation	Obtain and compare data from multiple sources and media, e.g. diary studies generate written, visual, and verbal data. Also compare different points of views on the same situation, gathered from different sample groups.
Analyst triangulation (Often referred to as *inter-rater reliability* in quantitative research)	Different team members collect, analyse, and interpret the data, as opposed to gathering data alone. Differently from the previous triangulation type, each team member collects and analyses data independently, then you all compare findings. In addition, all team members could first agree on analytic parameters or frameworks, then independently analyse the data, and finally compare findings.

Table 3.1 Types of triangulation to increase credibility of findings.

This helps you be more aware of their impact in the study. Document initial impressions and emerging patterns, which can be used later to inform data analysis.

- **Analyse negative cases.** Patterns or trends that don't fit within all participants or dominant themes act as exceptions to prove or broaden the rule. These help to find alternative explanations or see new perspectives.

- **Use thick descriptions.** Findings should be thoroughly reported, so that clients, team members, and even participants can empathize with the described experiences. Use participants' direct quotes and actual words to provide a window into their way of thinking and living. Chapters 7 and 8 expand on this point.

- **Describe prior similar work.** The examination of any prior work, supporting findings or providing evidence of other studies designed in similar ways, helps give the study credibility (Chapter 4). Similarly, the explicit mention of any familiarity gained with participants' cultural contexts in preparation for the study provides contextual information.

2. Can findings be transferred to another similar situation?

Transferability refers to the *external validation* of the study. It is associated with whether the study's findings have meaning for others with similar characteristics or in similar situations. When reading your study report or attending a presentation, stakeholders should be able to see a direct connection between the findings and another similar situation. As this is assessed by external people, your responsibility when reporting the findings is to be transparent and provide a clear description of the study. Chapter 4 describes each part of a field study that you will need to clarify when communicating findings.

3. Was the study conducted with care?

Dependability refers to whether *the process of the study* was consistent, reasonably stable over time, and methods were appropriately used. The description of the study should be transparent enough for a team member to examine the data and arrive at similar interpretations, findings, and conclusions. Any changes that occurred throughout the duration of the study and how these changes may have affected the way the study was conducted[13] should be clearly stated, as well as any issues during the process (e.g. was data systematically gathered in every session?) and the logic for selecting participants and using the methods (Chapter 4).

4. Are findings well supported?

Confirmability, refers to the *quality* of the findings and the degree to which they could be confirmed or corroborated by others. As with dependability, one way to address this criterion is asking a member of the team that hasn't been part of the study to act as "devil's advocate" and audit the study. They should question everything that has been done, from the appropriateness of the study design to the data collection and analysis methods.

5. How can findings be applicable to the project?

Applicability refers to how findings can be used to *inform the information design process* and ultimately the project. The impact and importance of the study and findings should be clearly articulated as specific recommendations, actionable items, or ways to move forward.

Revise these criteria before starting a field study. Some criteria should be taken into account since the beginning, when planning and designing the study (credibility), while others throughout the study, when analysing data (dependability, confirmability), and others need to be addressed towards the end, when reporting findings (transferability, confirmability, applicability).

---

13 Miles, M.B., Huberman, A.M., & Saldaña, J. (2013) *Qualitative Data Analysis*, SAGE.

How to work around constraints

While there are some real constraints for information designers, many are frequently exaggerated through lack of understanding of what is really needed to conduct a field study. So don't automatically rule out field research because of perceived constraints. Most common ones are *lack or little expertise*, *short time frames*, and *not enough resources*. The key is to know how to work around these constraints without losing essential aspects of qualitative research that otherwise can influence the quality of the study:

Expertise. If you have no or very little *research expertise*, the first time you conduct a field study, there will be an inevitable learning curve until you become familiar with research basics and fundamentals, but you must also start the study with the right mindset (Chapter 2). Each method poses different challenges but, with training, preparation, and practice, you can become good at using each of them; just don't expect to become a proficient information design researcher in one day. Read, test, and discuss your research ideas with colleagues before beginning an actual study and allocate some time to develop research sensibility.

Content expertise is also important. This is your baseline knowledge in the study topic and about your audience. This is most noticeable when working with a very specialized topic. In some cases, your unfamiliarity with the topic can be an asset, as you will ask the "dumb" questions that would be overlooked by someone more familiar with the topic. In other cases, this lack of knowledge can result in a failure to note or interpret important features of the study context. In both cases, the client plays a key role in helping you gain familiarity with the study topic. While you won't have to become an expert, you need some basic understanding, to ensure you cover all the key points in the design of your study and you speak the audience's language.

Time frames. In information design time frames are often short. Luckily, conducting a field study doesn't necessarily require spending four months or four weeks in the field. All methods can be adapted to shorter time frames. For example, in some cases, one or two days may be all you need to learn about your audience's commuting habits or observe them using a new design in their workspace. Most likely, you would adjust the study time frame to suit the overall project goals. Chapter 5 provides time frame estimates for each method.

Resources. A field study doesn't necessarily required big budgets because samples are relatively small, and a study can be conducted with few resources. Typically, most of the budget is allocated to travelling to participants' homes or workplaces and to participant compensations for their time. Furthermore, information design skills can be used to create promotional materials (e.g. to recruit participants) and design any other needed materials to gather data, saving considerable time and money if you had to externally commission these.

Thinking creatively in field research

In addition to research considerations and developing research sensibility, field research involves a significant portion of *creativity*. The creative part involves exploration, finding new angles, turning constraints into opportunities, risk taking, or designing engaging tools. With creativity, you can scale and customize any method to make it work under small budgets, short time frames, and limited resources. Here are some ways of adding creativity into your field study[14]:

- **Be open.** Have an open mind; be prepared to learn new things and accept multiple possibilities and ways of viewing the same situation.

- **Work with few resources.** It is very common in information design to think that time frames aren't optimal for conducting a field study. Make the most of the resources you have and look at the situation from different angles. There is usually a way.

- **Customize.** Interviews and observations are great methods, but you can make many modifications to ensure they are more engaging and relevant to your audience. You can tailor any method to the conditions and specifications of a particular problem and sample. These tweaks can help engage participants in the study and gather richer data.

- **Add rigour and precision to methods.** Often rigour and precision are seen as serious and boring, and less creative or playful qualities. A simple way to address both is to follow a systematic process, documenting every step. Find your own way of doing this by designing tools to make your work more enjoyable.

- **Deal with the unexpected.** A major part of field research is learning how to deal with ambiguity, as many components in a study are not predefined but flexible and constantly changing as the study progresses. Having a creative mindset can help make the most of these situations.

14 Patton, M.Q. (2002) *Qualitative Research & Evaluation Methods*, 3rd edn, Thousand Oaks, CA: SAGE, pp. 513–514; Blandford, A., Furniss, D., & Makri, S. (2016) Qualitative HCI research: Going behind the scenes, *Synthesis Lectures on Human-Centered Informatics*, 9(1), 1–115.

For example, how can you continue note-taking if you run out of paper, or turn an interview back on track when a participant is off topic?

Research methods have become very accessible for information designers in that there is a wide range of possibilities to choose from. Avoid only doing what you have learned in the past; always seek new knowledge. To know what is out there, go places, read books, watch movies, talk to different types of people; all these feed inspiration.

Use information design in the research process

Most qualitative data collected will be textual information. In this format, it is hard to see all you have because it is sequential and dispersed over many pages, rather than all displayed at the same time. This makes qualitative data hard to analyse by looking at two or three variables simultaneously or by comparing notes taken at different instances[15]. Building on Miles et al.'s concept of *visual displays*, I suggest that information design can be used to support the research process[16]. When raw data is given structure, it helps us make comparisons, notice differences, identify themes, count categories, see trends, and communicate findings with more clarity. Information design principles help minimize the risk of losing emotional tone, meaning, or richness from the qualitative data in the process of visual transformation. Throughout the next chapters, the book provides guidance for using information design aids in the following stages (Figure 3.3):

- **Mapping, exploring, planning, and gathering data:** Visualize initial understanding within a team to enhance early planning, and design materials and create templates to structure and facilitate data gathering [Chapters 4, 5, and 6].

- **Organizing and analysing data:** Prepare datasets in more accessible and engaging formats, and design visualizations to support sensemaking and reporting [Chapter 7].

- **Reporting findings:** Refine visualizations to communicate findings with clarity and create compelling stories [Chapter 8].

Information design aids created earlier in the process for data collection and sensemaking are rough and sketchy-looking. In some cases, these rough

---

15 Miles, M.B., Huberman, A. M., & Saldaña, J. (2013) *Qualitative Data Analysis*, SAGE; Babbie, E. (2010) *The Basics of Social Research*, 5th edn, Wadsworth Publishing.
16 Holtzblatt, K., & Beyer, H. (2014), Contextual Design: Evolved, *Synthesis Lectures on Human-Centered Informatics*, 7(4), 1–91.

INFORMATION DESIGN AIDS TO SUPPORT FIELD RESEARCH

CHAPTER 4: For mapping, exploring, and planning

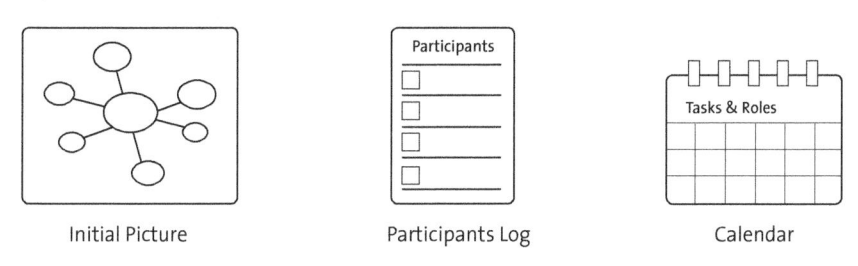

Initial Picture   Participants Log   Calendar

CHAPTERS 5 & 6: For gathering data

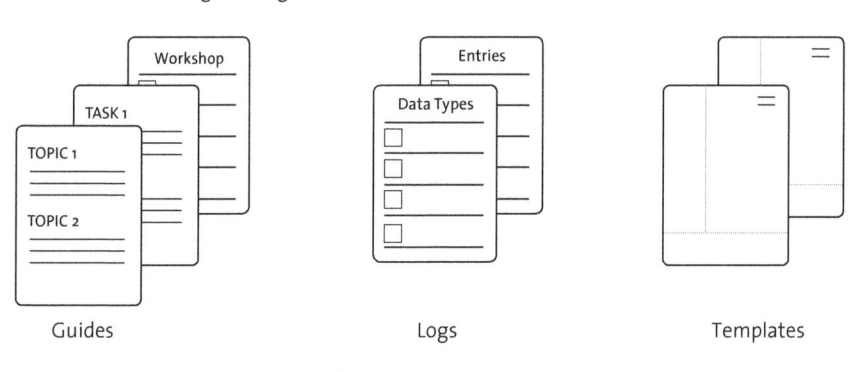

Guides   Logs   Templates

CHAPTER 7: For organizing and analysing data

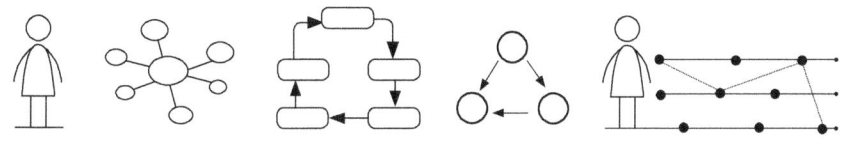

Rough Visualizations

CHAPTER 8: For reporting findings

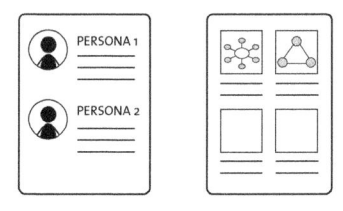

High-Quality Visualizations

Figure 3.3
Information design aids that can be used to support field research, discussed in the next chapters.

What is field research? | 51

visualizations are of large dimensions (bigger than notebook size), created by hand on a wall or table, and used as a tool to share data with the team. Those created towards the end of the process and included in a final report to present findings are of higher quality, often created using specialized software like Adobe Illustrator and InDesign.

Now you are ready to start planning and designing your field study. PART II introduces the research process and provides guidance to successfully complete each stage.

PART II:
Conducting a field study

4 How to plan and design a field study

You and your team have decided that conducting a field study is what you need for your project before moving forward to the next step or because your client has requested evidence to support one or many of your design decisions. The field research process introduced here presents a blueprint that you can use to plan and design the study. This process isn't always linear, and the steps are flexible: sometimes data collection and analysis occur simultaneously, and ways of collecting data change from one session to the next, based on initial learnings. This means the steps described in this chapter are indicative, rather than prescriptive, and should be taken as a guide or roadmap. Figure 4.1 provides an overview of the field research process, and questions within each step that you can use to better plan for study tasks and time frames. The following sections discuss each step in detail in the order you would most likely approach them.

Design the study

The first step is determining what type of study you need and what is the reason for conducting the study. As pointed out in Chapter 2, this will depend on the step of the information design process you are on and what you want to learn from the study. These are some possible reasons:

- Earlier in the process, you may want to explore how your audience follows existing instructions to put together new furniture to determine how they can be improved.

- Earlier in the process, you may want to examine the patient's journey in large hospitals to understand why people get frustrated and how you could improve the experience.

- Later in the process, you may want to evaluate whether your intended audience understands your new design concept.

- After you have launched a new design, you may want to examine whether your audience can do their job more efficiently when using your new interface or any changes should be made.

Once you have an initial direction of the type of field study you need and a clear reason, the next step is designing the study, which involves six broad activities:

1. Defining the goal: what you are trying to find out.
2. Determining the focus: what you will investigate to address the goal.
3. Selecting and recruiting participants: who you will gather data from.
4. Selecting methods for gathering data: how you will gather data.
5. Selecting methods for analysing data: how you will make sense of data.
6. Planning the communication of findings: how you will share learnings.

The following sections unpack each of these.

1. What you are trying to find out

Every field study needs an initial goal—what are you trying to learn?—even if conducted for exploratory purposes; otherwise, you won't know what to concentrate on during data gathering and analysis. The study goal is represented by a small number of *research questions*, which can change as the study progresses and your understanding increases. Typically, you would outline three or four research questions, each with sub-questions, adding granularity to what you want to know.

In some cases, to help you formulate research questions, you can create a picture (*conceptual framework*) to visualize your initial understanding and any assumptions about the study topic. This initial picture can take any form, such as a diagram or *mind map*[1], and explain what and who will be studied, the context, key areas, variables and how these are connected. Mind maps are helpful to show connections and thematic hierarchies between related ideas and concepts[2]. These maps present information, using an organizational structure that radiates from the centre, translating sequential information or seemly unrelated ideas into a memorable and highly organized structure. Mind maps consist of a main theme or topic, as a starting point or hub, from which associated sub-topics branch out; connections are shown using lines, symbols, words, colour, and images. Figure 4.2 shows examples of mind maps created to visualize initial understanding at the beginning of field research projects.

The picture will evolve as the study progresses and gathered data is added; the latest version could be used to guide the data analysis (Chapter 7).

---

1 Babbie, E. (2010) *The Basics of Social Research*, 5th edn, Wadsworth Publishing.
2 Buzan, T. & Buzan, B. (1996) *The Mind Map Book: How to Use Radiant Thinking to Maximize Your Brain's Untapped Potential*, London: Penguin.

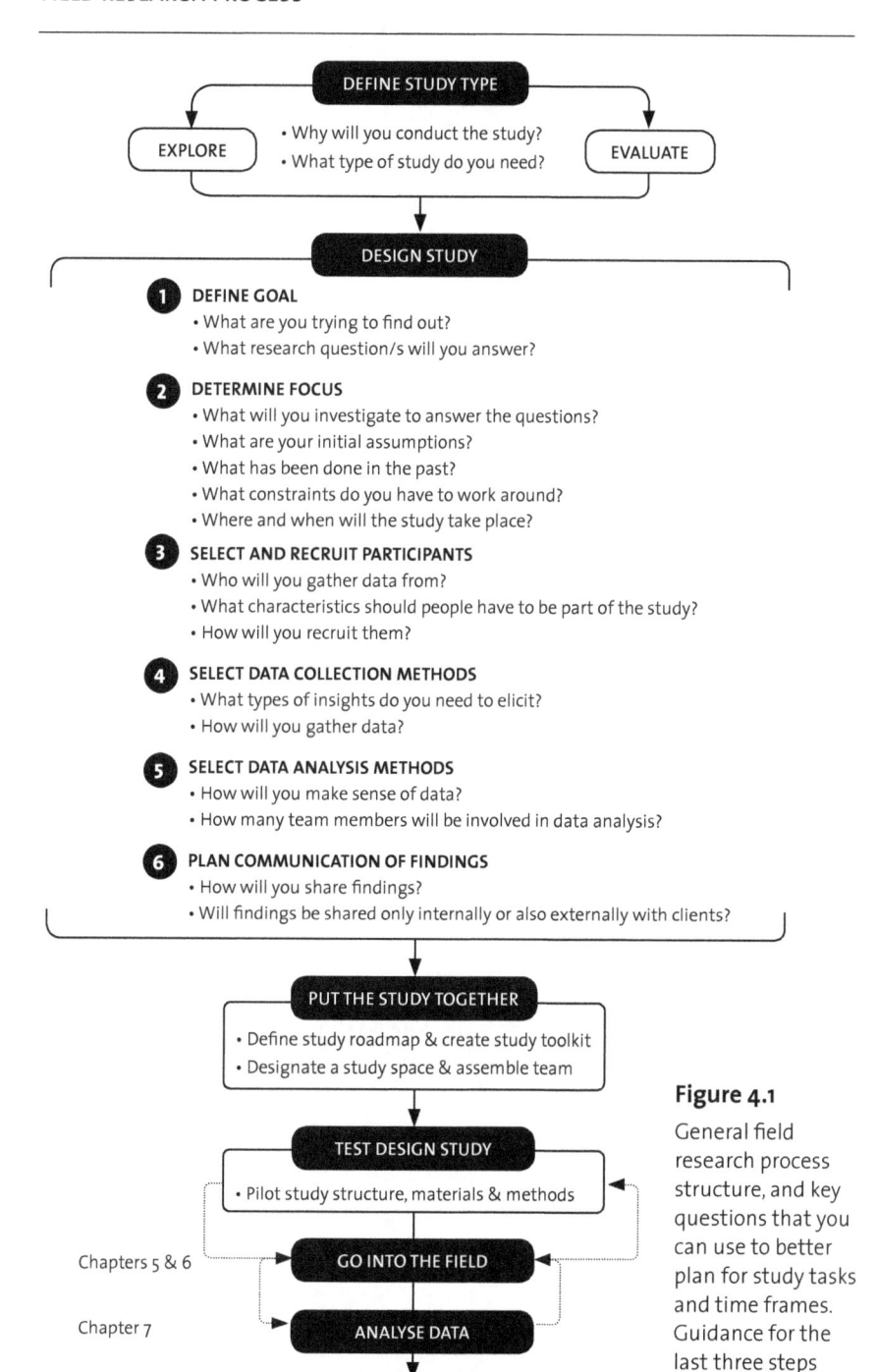

FIELD RESEARCH PROCESS

DEFINE STUDY TYPE
- Why will you conduct the study?
- What type of study do you need?

EXPLORE / EVALUATE

DESIGN STUDY

1. **DEFINE GOAL**
   - What are you trying to find out?
   - What research question/s will you answer?

2. **DETERMINE FOCUS**
   - What will you investigate to answer the questions?
   - What are your initial assumptions?
   - What has been done in the past?
   - What constraints do you have to work around?
   - Where and when will the study take place?

3. **SELECT AND RECRUIT PARTICIPANTS**
   - Who will you gather data from?
   - What characteristics should people have to be part of the study?
   - How will you recruit them?

4. **SELECT DATA COLLECTION METHODS**
   - What types of insights do you need to elicit?
   - How will you gather data?

5. **SELECT DATA ANALYSIS METHODS**
   - How will you make sense of data?
   - How many team members will be involved in data analysis?

6. **PLAN COMMUNICATION OF FINDINGS**
   - How will you share findings?
   - Will findings be shared only internally or also externally with clients?

PUT THE STUDY TOGETHER
- Define study roadmap & create study toolkit
- Designate a study space & assemble team

TEST DESIGN STUDY
- Pilot study structure, materials & methods

Chapters 5 & 6 — **GO INTO THE FIELD**

Chapter 7 — **ANALYSE DATA**

Chapter 8 — **SHARE FINDINGS**

Figure 4.1

General field research process structure, and key questions that you can use to better plan for study tasks and time frames. Guidance for the last three steps is provided in the following chapters.

How to plan and design the study | 57

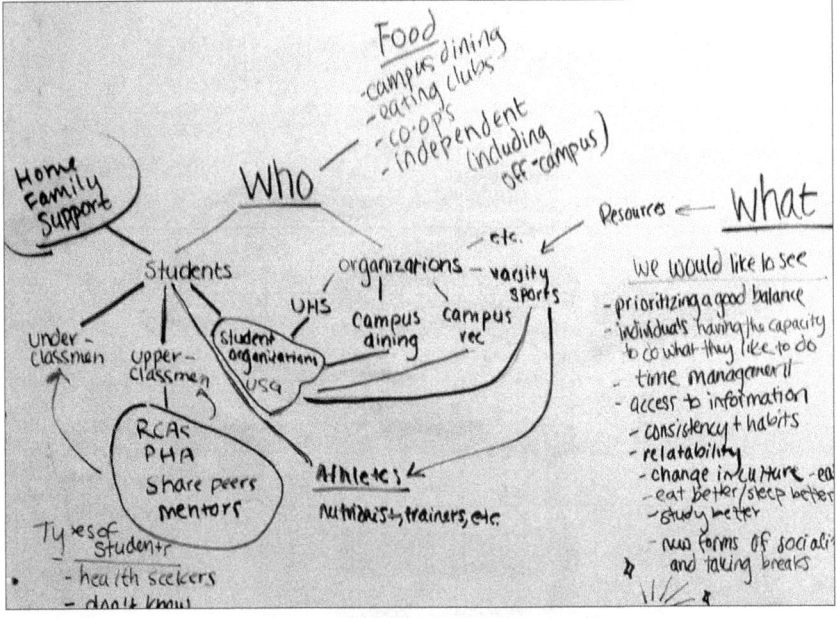

Figure 4.2

Mind maps. At the beginning of a field study, Princeton University students visualized as mind maps their initial understanding about the topic.

2. What you will investigate to address the goal

After defining the study's goal, the next step is to determine the *unit of analysis* or specific focus. The focus is what you will be primarily paying attention to when going into the field to answer your research questions. The focus varies from project to project[3] and can be defined by almost anything that could provide the understanding you need, such as a role (CEOs, teachers), a small group of people, individuals, families, cities, an event, an organization, people who share a common experience (e.g. commuters), tools (e.g. a map), an environment (e.g. a museum), a period of time (e.g. a day in the life of...), a process (e.g. filing tax returns), a culture, etc. Other dimensions you can define to have a more specific focus are its social size (individual, community, culture, nation), its physical location (a specific organization, public places, etc.), and its temporal extent (daily, seasonal, etc.)[4].

Defining the unit of analysis will help you determine the location, possible time slots, and the initial number of sessions that the study will need to gather enough data to address the goal. A field study can involve only one long session with one participant or multiple sessions with different participants. Each encounter with a participant constitutes a session and, depending on the study, you can meet one or more times with each participant. For example, if your focus will be on a specific role, like police officers, you will have to clearly delineate who would be a potential participant, when you could conduct the study, how you can recruit them, and whether it would be more insightful to conduct many sessions with one police officer or just one session with different officers. But if your focus will be on a specific environment, like a supermarket, from a specific neighbourhood, you will have to determine what would be the best time slots to gather the data you need. The number of sessions will also depend on the data collection method of the study.

Check resources and identify constraints. Once you have a better sense of the study's goal and focus, determine what resources you have available for conducting the field study and identify what constraints (time, budget, expertise, resources) you must work around. For example, if you have an extremely tight time frame, you could scale down the initial goal or reframe the research questions or just focus on one of the research questions rather than on all of them. Or if getting access to the initial unit of analysis would be nearly impossible (e.g. director of the hospital) because they would be out of the country for six months, identify what other related units of analysis (e.g. deputy director, other department directors, hospital colleagues) could help you learn about them.

---

3 Patton, M.Q. (2002) *Qualitative Research & Evaluation Methods*, 3rd edn, Thousand Oaks, CA: SAGE.
4 Miles, M.B., Huberman, A.M., & Saldaña, J. (2013) *Qualitative Data Analysis*, SAGE.

Review what has been done in the past. In parallel, gain a better understanding of the topic and unit of analysis you will investigate. This understanding is mostly gathered through *desk research* or a *literature review*, often complemented by *expert consultations* (about content, people, logistics) and discussions with members of the organization related to the problem at hand. To some extent, data collection starts here: the minute you decide to plan a field study and before conducting the actual study. Content generated during client conversations, preparative meetings with the team, and any project-related documents like presentation decks and reports, constitute important *primary data* that helps shape the study and refine its goal.

These initial learnings and having a clear focus give direction to desk research and determine which case studies or other resources to examine first. Particularly, the review of national and international case studies can be very eye-opening, as well as previous relevant projects and cases from other industries that seem to be directly related and those that may be unrelated. For example, if you were to focus on learning about the patient's journey, you could look for prior research studies done in hospitals with a similar goal. This will give you baseline knowledge about core steps in a patient's journey or when would be best to go into the field to learn the most or who would be the most appropriate people to talk to. Regardless of what you review, your aim should be to:

- Gain a thorough understanding of your unit/s of analysis.
- Understand how they have been investigated before and why.
- Understand whether prior similar studies have been successful or failed and, in either case, why.
- Gain familiarity with who are or may be your participants, and discover whether you first need to dig deeper into specific areas or topics.
- Determine who should be involved in the study.
- Gather evidence to clearly explain to clients and stakeholders why this type of field study is needed before moving forward in the information design process.

This knowledge helps identify basic but frequently underestimated things to consider when planning a field study such as determining key characteristics for potential participants or defining appropriate study settings. In addition, pay attention to those things which haven't been adequately addressed in previous studies or didn't work as planned. The more thorough your review, the better equipped you are to move forward designing the study.

3. Who you will gather data from

Based on your study's goal and focus, determine the kind of people you want as participants of the study by deciding the specific characteristics that potential participants should have: the *eligibility criteria*. The next task is to decide how many participants you need: the *sample size*. The final task is deciding how to recruit them. This involves determining a *sampling strategy* and choosing a *recruitment method*.

Eligibility criteria. These are the specific characteristics that people should have to qualify as participants, which vary from one project to another because they are based on the study's goal. In some cases, these can be quite broad (e.g. 20–45-year-old people who commute by bus), in others, very focused, resulting in a specific type of sample (e.g. 20–45-year-old female nurses who commute by bus). The more specific and narrow the eligibility criteria, the harder to find people who qualify as potential participants. Regardless of the type of field study, it is essential to delineate who the sample is. Furthermore, as much as possible, work with participants who represent the intended audience; for example, it wouldn't be appropriate to recruit only residents of a city if the goal of the study is to better understand how a new museum map works as tourists may also be a large part of the audience (Table 4.1).

Sample size. Field research works with relatively small samples of participants. There aren't rules for sample size; it depends on the study's goal, what you want to learn, how findings will be used, and, most importantly, what is doable with the available time and resources. It is preferable to recruit just a few participants who fully respond to the eligibility criteria and from whom you can gain rich insights, rather than recruiting a larger sample but not gathering useful data. Seek depth, by spending considerable quality time with each participant, rather than breadth. Some studies could involve only one or two in-depth cases (e.g. a day in the life of a CEO).

Sampling strategy. This can be used to recruit participants, based on the eligibility criteria[5]. Of the many recruitment techniques, a few, relevant for field studies in professional contexts, are discussed here:

- **Convenience sampling:** This technique involves working with the most accessible participants, even if they don't satisfy all eligibility criteria: possibly colleagues or friends. This technique is the most convenient to save time, money, and effort, but resulting insights may not be fully representative.

5 ibid.

	ELIGIBILITY CRITERIA	SAMPLE SIZE	SAMPLING STRATEGY	RECRUITING METHOD	STUDY LOCATIONS
CASE STUDY 1: The Redesign of the Carnegie Library of Pittsburgh	• 18 years of age or older • Users and staff working in the library • All genders	< 30	• Purposeful sampling	• Direct contact	• Carnegie Library of Pittsburgh
CASE STUDY 2: Legible London	• Residents and tourists commuting in London by tube or by foot • All genders	< 400	• Purposeful sampling • Convenience sampling	• Direct contact • Advertising	• Tube stations • On the streets
CASE STUDY 3: Vendor Power Guide	• 18 years of age or older • Street vendor population • Working in New York City • All genders	30	• Purposeful sampling	• Direct contact	• On the streets • Vendor meetings
CASE STUDY 4: A Better A&E	• Visitors, patients, and staff working at the A&E departments in specific UK hospitals • All genders	<60	• Snowball sampling • Purposeful sampling	• Direct contact	• Hospitals
CASE STUDY 5: To Park or Not to Park	• 18 years of age or older • Driving a car • Living in New York City • All genders	7	• Purposeful sampling	• Direct contact • Advertising	• On the streets

Table 4.1 Participants' field study details for the case studies discussed in PART IV.

- **Purposeful sampling:** This technique involves selecting participants who fully satisfy all eligibility criteria and, thus, are most likely to address the goal of the study efficiently.

- **Snowball sampling:** This involves finding participants by asking the first recruited participants if they know anyone who satisfies eligibility criteria and would be interested in the study. This is a useful technique for accessing hard-to-reach populations, such as CEOs of big organizations.

- **Opportunistic sampling:** Differing from the first technique, this involves finding participants on the spot while doing something else, such as attending an event or a client meeting, even if they don't satisfy all eligibility criteria.

Extreme cases, participants who don't respond to the eligibility criteria but are indirectly related to the study's topic, can provide a very different and unique perspective to the situation that more familiar participants won't have. Depending on each project, it is a good idea to gather data from one or two of these cases.

Recruiting method. The final task, after choosing a sampling strategy, is to decide how to reach out to potential participants. These are three possible ways[6]:

- **Direct contact:** Approach potential participants in the workplace (with the necessary authorization if needed) or in public spaces[7]. Or use the Internet and social media to reach out to potential participants.

- **Advertising:** Advertise on noticeboards in physical spaces, subject pools, through targeted email lists, via online lists or social media.

- **By third parties:** An intermediary within the organization contacts potential participants. You should provide clear guidance to explain the eligibility criteria and share any ideas about the ideal profiles of participants.

Participants' attitudes. Curiosity, opportunities to learn, the study's goal, or having fun are influencing factors that determine people's willingness to participate. If you are testing a new solution, people may often be inclined to participate to learn about the idea first. In other cases, if you are conducting a study in your client's organization or company, people are most likely going to participate, even without anything in return, but compensation is almost always necessary when recruiting random participants (Chapter 3).

4. How you will gather data

Simultaneously, while deciding whom your participants could be, you also need to start thinking about how you will gather data from them in the field. Qualitative methods for data collection help access people's past, present, and ideal future experiences and stories by examining what they say, do, think, or create[8]. Some methods are more appropriate for focusing on past experiences, while others suit the present and future; and other methods are more appropriate to elicit a specific type of knowledge but not another.

To help select the right data collection method, it is useful to become familiar with how people process information and store memories—that is, the different types of memory and the types of knowledge they store—and how to access those memories. Equally important is knowing what type of knowledge is stored in each memory and how to access it.

---

6 Patton, M.Q. (2002) *Qualitative Research & Evaluation Methods*, 3rd edn, Thousand Oaks, CA: SAGE; Blandford, A., Furniss, D., & Makri, S. (2016) Qualitative HCI research: Going behind the scenes, *Synthesis Lectures on Human-Centered Informatics*, 9(1), 1–115.
7 Always take into account your safety and gathering informed consent.
8 Sanders, E.B.N. (2002) From user-centered to participatory design approaches, *Design and the Social Sciences: Making Connections*, 1(8); Sleeswijk Visser, F. (2009) *Bringing the everyday life of people into design*, PhD thesis, Delft University.

Types of memories. Information is stored in the brain in the form of knowledge and in different types of memory: short- and long-term. Short-term memory stores information for just a few seconds and is of limited capacity (e.g. a new ATM password); distractions lead to the information stored there being lost forever. For example, once you complete a task (e.g. enter new password), you won't remember the details of what you have just done. In contrast, long-term memory stores indefinite amounts of information for years, which is relatively easy to access; for example you can access by asking questions about what people remember. In contrast, short-term memory is much harder to access because it is ephemeral. There is a variety of methods you can use to access each memory depending on the type of knowledge you need to elicit.

Types of knowledge. Broadly, information stored in long-term memory is *explicit* and *implicit* knowledge. *Explicit knowledge* involves the memories and information that people have at the top of their minds and can express with words. This type of knowledge is easy to retrieve as it is mainly manifested by **what people say** when responding to direct questions. Often it also represents what they want us to hear, but people tend to filter personal details or omit information they don't consider relevant. This type of knowledge is accessible by most data collection methods because it is readily available. Examples of this knowledge are your name, date of birth, your parents' names, or your address.

Implicit knowledge is harder to access and articulate but not as difficult as that stored in short-term memory, and it can be accessed by looking at **what people do**. These elicitation methods focus on identifying people's behaviours, attitudes, and preferences. You can observe what they wear, how they interact with other people, or the steps they performed to achieve a goal or task. However, if you want to understand why they do what they do, or elicit a deeper level of detail beyond what you can see, you need to access knowledge stored in their short-term memory.

Information stored in short-term memory is *tacit knowledge*. This is the knowledge involved in ideas, skills, and experiences that people use but they can't articulate or express how they use it[9]. To elicit this type of knowledge you need to examine **what people think** and **how they make decisions** often using a combination of methods[10]. These methods focus on examining expert behaviour and helping people remember exactly how they accomplished a task they are extremely used to doing because the more we practise a task, the less we explicitly think about it when performing it. A clear example could be asking a seasoned chef to explain the steps they fol-

---

9 Polanyi, M. (1983) *The Tacit Dimension*, Gloucester, MA: Peter Smith.
10 Petre, M. & Rugg, G. (2007) *A Gentle Guide to Research Methods*, Open University Press, McGraw-Hill.

low to bake a chocolate cake. Most likely they won't recall the exact order in which they perform each action because the task no longer requires conscious thought. To elicit tacit knowledge, you can observe what people do, while simultaneously asking questions because practical demonstrations (e.g. how to perform a task or use a product) reveal many steps and nuances absent from verbal accounts. Another way to access this knowledge is asking people to think aloud and verbalize their immediate thoughts while performing a specific task. This forces people to explicitly think about what they are doing, describe each step, and articulate details[11]; otherwise, people take obvious details for granted, forgetting to mention them. Also, you can show people a visual aid to help them recall any missing details.

To access tacit knowledge you can also examine **what people wish for or their ideals**. Often asking direct questions about predicting their future may not be the most appropriate approach because people may feel blocked or genuinely never have thought about it. More appropriate methods focused on eliciting this knowledge ask people to make things or draw, as a way to express these types of thoughts, feelings, and dreams. For example, if you ask people to create their ideal way to communicate with their loved ones, you are asking them to reconnect with their latent needs (Chapter 1), which is another form of tacit knowledge.

So, before selecting what methods you need for your study, discuss with your team and identify the types of insights you need to elicit to better inform your design decisions and move forward in the process (Figure 4.3): should the study focus on identifying needs and behaviours? Or should it focus on identifying specific steps in a process? Or should it focus on learning participants' ideal solution? Shedding clarity on the type of insights you need, will help you determine what is the most appropriate method. For example, conducting an observation study during the early stages of a project could help you learn how people use an interface that you are planning to redesign; that is, behaviours. These insights would help you identify what aspects of the existing interface could be improved to better support their behaviours. However, if you are working on an undefined project to help students better manage information, it may be more helpful to ask that population to imagine and create their ideal ways of dealing with class content and reading materials; that is, latent needs and wishes. These insights would help you generate solution ideas more in line with what students need.

The next two chapters present a selection of methods that will help you obtain the holistic understanding you need to empathize with your intended audience: observational studies, contextual interviews, contextual inquiry,

11 ibid.

diary studies and design probes, and collaborative workshops. Building on Sanders's[12] classification of techniques, these methods focus on gathering data about *what people say* (opinions), *what people do* (behaviour), *what people think* (decisions), or *what people create* (ideals).

Most of these methods can be used alone or in combination for exploratory or evaluation studies by changing their focus and tailoring them to the specifications of each project and stage in the information design process. Chapter 5 describes how to use these methods for exploratory studies, and Chapter 6 explains how to use them for evaluation studies.

5. How you will make sense of data

You won't start the analysis step until you have gone into the field and gathered some data. However, it is useful to think in advance and coordinate with the team (if you are working with one) how you will approach this part of the research process because the analysis of qualitative data involves patience, time, work, and a high degree of interpretation by finding patterns, connecting apparently unrelated data, and holistic thinking. Making sense of this type of data also requires following a structured and systematic way of thinking and working, and not skipping steps (even when the time frame seems too tight!).

There are some decisions you can make at this early stage in the process to make the data analysis step more organized and smoother. Some of them are: will data be analysed by one person or many team members? How can we ensure that all members will be analysing the data following similar criteria? How will we be analysing the data: manually or using software? If the latter, do we have it and know how to work with it? How can we keep data gathered organized and accessible to everyone in the team during the analysis? Chapter 7 provides further guidance for the analysis process and directions for using a selection of methods to support the process.

6. How you will share learnings

In most cases, you will conduct a field study to inform your information design process and your team will be the only party to whom you communicate findings and insights. However, in some cases you may have to report back key findings to clients; for example as an evaluation report, executive summary, or meeting presentation. At this stage in the research process it may be too early to define how a report would look like, but it is important to determine beforehand whether findings will be used only internally or will

---

12 Sanders, E.B.N. (2002) From user-centered to participatory design approaches, Chapter 1 in Frascara, J. (ed.) *Design and the Social Sciences: Making Connections*, London: Routledge.

TYPE OF METHOD	TYPE OF KNOWLEDGE & INSIGHTS	
• Interviews	**What people say** Explicit Knowledge EXAMPLES: • Opinions • Facts • Stories • Descriptions • Attitudes	EXPLICIT NEEDS
• Observations	**What people do** Implicit Knowledge EXAMPLES: • Expressions • Interactions • Behaviours • Dynamics • Observable needs	
COMBINATION OF METHODS: • Observations and interviews • Think aloud • Describe visual aid	**What people think + How people make decisions** Tacit Knowledge EXAMPLES: • Steps • Processes • Cycles • Frameworks • Sequences • Hierarchies	
• Making and drawing • Other generative methods	**What people create** Tacit Knowledge EXAMPLES: • Dreams • Wishes • Latent needs • Ideals	DEEPER KNOWLEDGE IMPLICT NEEDS

INFORMATION DESIGN RESEARCHER   STUDY PARTICIPANTS

Figure 4.3

Suggested methods to elicit specific types of knowledge and insights.

How to plan and design the study | 67

also be shared with other parties (e.g. clients, stakeholders, or participants). Chapter 8 discusses ways to communicate findings internally and externally.

Assemble the team

Depending on the project and the size of your team, you may have to conduct the study solo or decide to do it with team members. In each field session, the goal is to capture what you see and hear by taking notes. Note-taking can be really difficult when done in tandem with looking, asking questions, and active listening. This is why when possible, field research is easier done at least in pairs, so you can divide tasks, assign roles, and discuss points of view throughout the process. If you are going into the field in pairs, define-clear roles beforehand; for example, in an interview session, if one person has the role of asking questions, the other is the note taker.

To avoid becoming overwhelmed during data collection, and to facilitate data analysis, it is essential to start the process with good preparation. If many team members will go into the field and take notes, determine a baseline and agree on which tools will be used to record each session. This will help ensure consistency throughout the data-gathering process as everyone will follow a similar format, and gather data of similar quality and depth.

Put the pieces of the study together

After you have a clearer understanding of what your field study will involve, the next step involves acting on those decisions and start making the study a reality. This involves putting together a study roadmap, creating the study toolkit to support your work in the field, and designating a study space.

Study roadmap

The *study roadmap* (or research protocol, Figure 4.4) describes all decisions made about the study during the early steps of the research process and it gets populated as the study progresses. In the same way you visualized your initial understanding of the study topic, creating a document describing the study's initial plan and structure helps produce a better sense of what needs to happen and when. This document can take any form, from a traditional written report to a more schematic representation or map. In any case, it should include the study's goal, the focus, sample recruitment strategies, methods selected, materials that need to be designed or ordered, and any

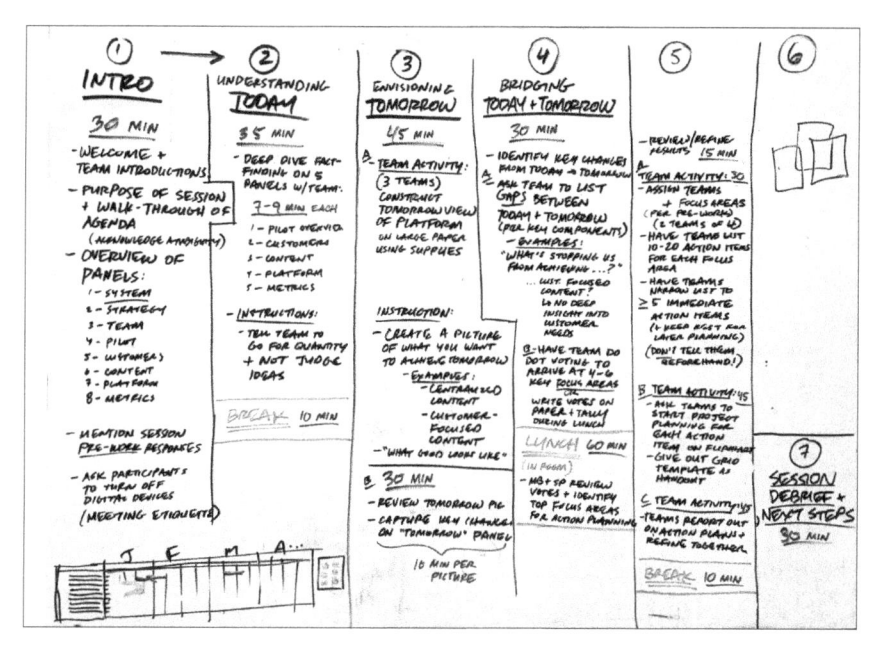

Figure 4.4
Study roadmap to plan a workshop session.

other relevant details that have been decided so far (study settings, days of the study, etc.).

As part of the roadmap, you can create a *timeline* or *calendar* (Figure 4.5), similar to a Gantt chart, as a way to show any tasks scheduled over time[13]. You can create this in Microsoft Excel or other project management tool, but if working with a team, it is useful to create something of large dimensions so everyone can easily access any study-related information and follow the progress. This could be a wall timeline, indicating key tasks, assigned roles, and due dates aligned to the overarching project goals and time frames. If working solo, having a clear, initial schedule helps manage time more effectively.

Planning and creating a shared understanding of the study among team members helps to keep everyone involved in the study on the same page and to considerably reduce data collection and data analysis times. Even when working under tight time constraints, don't rush or skip this step, and be open to changes as the study progresses.

---

13 Kirk, A. (2012) *Data Visualization: A Successful Design Process*, Birmingham: Packt Publishing.

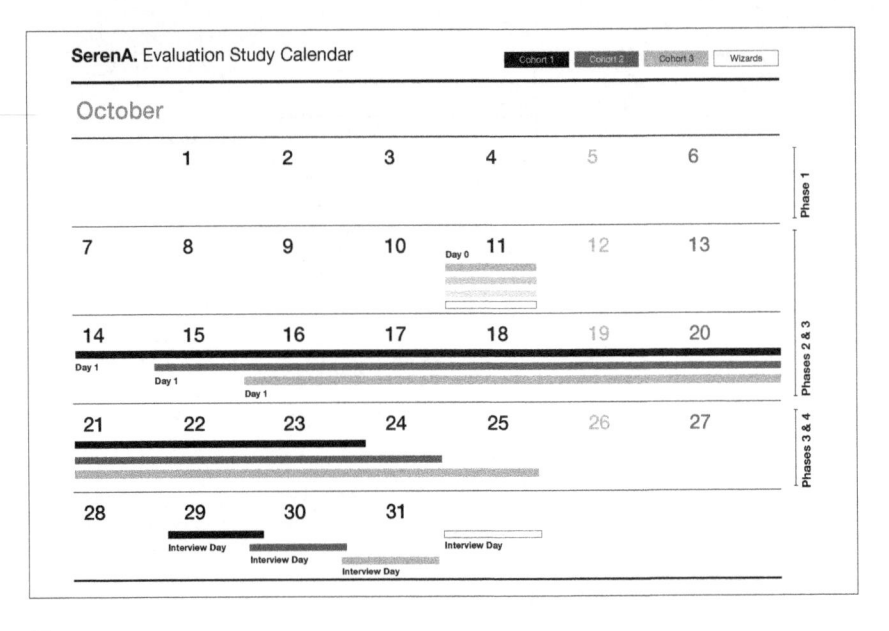

Figure 4.5
Study calendar. Time frames for the SerenA study conducted at University College London to indicate dates for training sessions (Phase 1) and interviews (Phases 2 and 4), and durations of each self-documentation period for each participant group (Phase 3).

Study toolkit

Based on the study roadmap, you will have identified a range of communication materials and supporting instruments that you can create, and tools you need to aid data elicitation in the field and other activities in the research process (Figure 4.6). These will form your study toolkit.

Materials. These comprise everything you need to advertise the study to recruit participants, such as digital or print posters, flyers, invitations, emails; to customize the study to your participants' likes, such as branding elements, thank-you cards; and to provide participants with information about the study, such as consent forms and information letters.

Instruments. These are specific aids to help give the study structure and initial direction, and elicit the necessary data in a systematic and organized manner. Some of these are: guides, templates, logs, checklists, heuristics, briefs explaining tasks and activities, instructions, and reminder postcards. As the first three are common to most methods, here is some guidance of what they should look like:

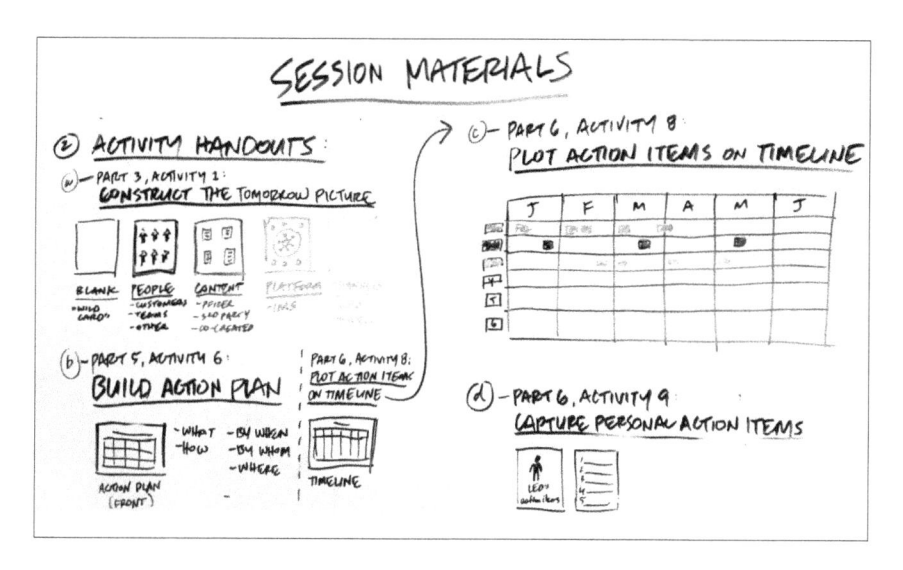

Figure 4.6
Representation of materials and instruments needed for a workshop session.

- **Guides** are short documents with a list of specifications, topics, or questions, organized by themes or priority. A guide isn't a script which presents questions that must be asked to each participant in a session; rather a guide is a list of key words and topics regarding what you would like to learn from each session (e.g. specific interactions, actions, tasks) (Figure 4.7). They can be used as checklists to ensure that observational objectives are completed or questions on all key topics asked. Guides can take many forms, from a simple list of questions to aid memory to also including space for note-taking. The content of a guide often changes from one session to the other based on initial learnings. You will create guides for observational and interview studies. But you can also create a Recruiting guide to have a better sense of who you should recruit for the study.
- **Templates** are documents created to capture data in an organized way. Each template can include a label, indicating the type of note it should be used for—e.g. direct observation or interview answer— space to write participant's name, and any relevant details about the session (e.g. date, time and setting). Templates for observational studies can also include spaces to describe sequential structures, durations, temporal locations, and contradictions. Templates don't have a strict layout; they can be designed in many ways (Figure 4.8). You can create templates for almost any method.

How to plan and design the study | 71

INTERVIEW GUIDE EXAMPLE

INTRODUCTION
- Explain goal of study
- Explain NDA (if applicable)

WARM-UP QUESTIONS
- Request permission to record session, data consent, and take photographs
- Ask for something not related to the study to learn more about the person (e.g. have a tour of the office)

CORE IN-DEPTH QUESTIONS
THEME/TOPIC 1
1. Possible areas to ask and two or three starting questions:
   - Participants' understanding about the topic, their process, their criteria, tools they use, their struggles
2. Activities (if applicable)

THEME/TOPIC 2
1. Possible areas to ask and two or three starting questions:
   - Participants' understanding about the topic, their process, their criteria, tools they use, their struggles
2. Activities (if applicable)

CLOSURE
- Wrap up and give compensation
- Review of photos taken and take more if needed
- Discuss follow-up steps (if applicable)

Figure 4.7
Example of an interview guide to explore two different themes or topics.

- **Logs** are tables for managing and documenting participants' demographic information, the types of data collected from each of them, and when each session takes place. They can also be used to document more dimensions of a study, such as whether all participants are asked the same questions or monitoring who misses an entry in a self-documentation study (Figure 4.9). You can create logs using Microsoft Excel, Adobe software, or by hand. They can be as big or small as needed, but, if working with a team, or another person will also gather data, large-scale sizes are preferable, so everyone can contribute to the log. As the study progresses templates help populate a log for each participant.

None of these instruments is prescriptive. All can be tailored to the needs of a study and altered as the study progresses.

Tools. You also need tools for capturing and recording data in the field. These could be digital or analogue and include notebooks, eye-tracking equipment, index cards, journals, sticky notes, coloured pencils, markers, whiteboards, highlighter pens, audio recording (e.g. iPhones, iPads), photographic cam-

INTERVIEW AND OBSERVATION TEMPLATE EXAMPLES

INTERVIEW NOTES TEMPLATE
Date & Time:
Participant:
Location:
General notes
Core notes
Theme/Topic 1
Activities (if applicable)
Theme/Topic 2
Activities (if applicable)
Emerging questions
Closure & follow-up notes
Take photos of the place or create sketches!

OBSERVATION NOTES TEMPLATE
Date & Time:
Participant/s (if applicable):
Location:
PART 1 Notes: What you see, hear, smell, taste   **Descriptive observations**   Notes about space, people, characteristics, interactions
Focused observations   Notes about specific person or group, problems, objects, interactions
Selective observations   Notes about main person problem, object, interaction
PART 2 Notes: Analysis of what you learned
Take photos of the place or create sketches!

Figure 4.8

Examples of templates for observational studies and interviews including spaces for note-taking under key sections. Other templates may include only one big space for note-taking.

era (e.g. disposable camera), video camera, screen capture software, digital or traditional self-documentation kits (e.g. diaries, postcards, tasks), laptop, or iPad. Larger studies can require specialized software to analyse collected data (e.g. NVivo, MaxQDA, Dedoose, or ATLAS.ti) (Chapter 7). Online studies can require digital tools to enable different forms of online communication like email, mailing lists, and blogs.

In addition, you can develop an abbreviation system or coding for note-taking. This helps optimize data collection, facilitate data access during analysis, and share data with the team. For example, use a different colour pen to distinguish different types of notes while a participant is talking, another colour to keep track of what questions you asked, quotation marks to indicate key participants' exact words you want to remember later, and indicate contradictions between brackets.

PARTICIPANT LOG TEMPLATE EXAMPLES

PARTICIPANTS' SESSION OVERVIEW–SINGLE METHOD STUDY						
	P1	P2	P3	P4	P5	P6
Demographics						
Role						
Session type						
Location						
Length						
Key emotions						
Quote						
Key values						
Quote						
Key behaviours						
Quote						
Key struggles						
Quote						
Follow-up						

PARTICIPANTS' STUDY OVERVIEW (Data shown below is illustrative only)						
TYPE OF DATA + FORMAT	P1	P2	P3	P4	P5	P6
Interview 1	7/7/18	7/7/18	8/7/18	8/7/18	9/7/18	9/7/18
Transcripts	yes	N/A	yes	yes	N/A	yes
Interview 2	10/7/18	N/A	10/7/18	N/A	N/A	N/A
Observation 1	TBD	10/7/18	N/A	10/7/18	N/A	TBD
Observation 2	N/A	TBD	N/A	TBD	N/A	N/A
Documents	yes	yes	N/A	yes	yes	N/A
Workshop	12/7/18	12/7/18	12/7/18	N/A	N/A	12/7/18

Figure 4.9
Use the first log to help keep participants' data organized and capture key initial findings and first impressions from each session for each participant.
Use the second log if you will conduct multiple sessions with a same participant and/or use different methods. It helps keep track of when and what type of sessions have been conducted with each participant and when the next sessions are scheduled.

It is important to plan what materials, instruments, and tools you may need at each step of the research process as it may take time to design and test the appropriate materials and instruments before the study begins. Also, specific software or tools may take time to find or be delivered.

Study workspace

If possible, allocate a physical space—such as a meeting room—for the duration of the study. Having a dedicated space is particularly relevant when working with a team, because everyone can document progress maintaining a shared understanding. There is no right way to use this space, but having a shared understanding of how to organize study materials helps keep everyone on the same page. For example, you can dedicate each wall to a different purpose: a planning wall to show logistic information such as study timelines and calendars, a data wall to capture incoming data, an analysis wall to indicate connections between data and participants, and a concept wall to indicate emerging ideas or concepts. The space should also be big enough that you and your team can discuss experiences in the field, and, later in the process, contribute to the data analysis.

Test the design study

The final step before going into the field is checking whether your plan (research protocol and time frame), what you have designed (materials and instruments), and your chosen data-gathering methods will actually help you to gather data—or gather the appropriate data. This is called *piloting* and can be done at different steps in the process. Before starting data collection, piloting sheds light on how long a participant would take to complete each task, or how much data is needed. Later in the process, piloting is useful to determine the best way to make sense of collected data, or how long the analysis of a day of interview data would take.

There are many ways of piloting. Simpler pilot studies involve testing any materials, instruments, and tools created for the study. Others involve running a short version of the study with one or two participants early on, to check the flow and identify unclear areas. Consulting an external researcher to check whether your plans are feasible is yet another.

Piloting provides insights to identify flaws and make any major reviews or minor changes, before starting the main study. Time spent in piloting helps plan the study more effectively and actually saves time and money overall, as you minimize the risk of collecting useless data or asking the wrong questions. Piloting ensures that you gather high-quality data during the actual study.

Going into the field

The purpose and goal of the study has been defined, and the focused specified. You have a study structure and decided which methods are more appropriate to use to gather the type of insights you need. Now it is the time to go into the field and take notes.

Understand note-taking in the field

Regardless of the method you will use for the study, when you are in the field your goal is to capture what you see as accurately as possible, what each participant generates, and what participants say by writing, drawing, or recording what you see and hear. Also document what you feel and think about each session. As a general rule, write notes as soon as possible; that is, while collecting data or right after you finish a session, but always write notes before the end of the day you gathered the data, to avoid forgetting important details. The more sessions you conduct in a day, the harder it will be for you to remember all interesting data points. Figure 4.10 provides examples of different types of field notes.

What you see and hear. Field notes should be descriptive, authentic, concrete, and detailed so they reflect the context within which people interact[14]. Include:

- date, time, location of each session
- who was present
- a description of the setting (smells, sounds, textures)
- specific characteristics about each participant
- any social interactions and activities that occurred

Also, focus on capturing key words, expressions, gestures, tones, accents, and as many details as possible, but don't include generic terms to describe specific instances or activities such as "anger", "poor", or "nice". Adjectives and nouns alone aren't descriptive, and have little meaning for someone who wasn't present in the setting. Field notes should include enough details so that later when you or another team member reads them is able to understand the observed setting.

Include participants' responses to specific questions in the form of direct quotations and using exact words and phrases. Also include lists of major points made by participants, key terms or words, and specifics such as

---

[14] Patton, M.Q. (2002) *Qualitative Research & Evaluation Methods*, 3rd edn, Thousand Oaks, CA: SAGE.

EXAMPLES OF FIELD NOTES

Notes about what you see and hear

Monday, January 15. Client's office, 10th floor, 8:30am. Medium size female in mid-30s. She wears jeans and a blazer. She picks up the phone and discusses project time frame: *"When will the sales department have the presentation ready?"* She writes notes on post-it while speaking in the phone. *"But Paul, we need that presentation ready by tomorrow 9am."* She nods and says: *"Thank you, I appreciate your effort"*. And hangs up. 4:10 pm.

Notes about what you feel

I came to Karen's office bright and early as she requested, but she didn't come in until almost half an hour later. Initially, the wait made me feel a bit annoyed. Now I'm trying to leave that feeling aside.

Notes about what you think

Karen seems concerned today, but also seems relieved that the work will be done in time. I think she writes down notes to calm her down, but she is also appreciative of what others are doing and lets them down.

- Not really making eye contact
- Was sitting very formally as if I was interviewing her for a job

Seemed nervous, so I am unsure if her responses were legitimate or fabricated/biased.

Interview response

"My family was not that physically active at all, I was encouraged to do sports sporadically. I was more involved in musical theater as a kid. Growing up my mom did not cook super healthily, stuff like mac n cheese, spaghetti, we always had vegetables though. The past couples years she has gotten into more healthy cooking habits. My dad also smoked growing up. My parents didn't really model healthy habits. I didn't start eating healthy and exercising until I left and went to boarding school in 9th grade".

Diary study or design probe notes

"I was a bit hesitant to just flip the first card and get started; I wanted to get a feel for how the cards would shape or interact with my normal workflow. As a result, my only use of the cards so far has been an exercise in familiarity. The effort was worthwhile, the cards no longer feel foreign or mysterious".

"Today was a bit of a strange day, because we had a team meeting external with the whole business service department. I had the meeting in the morning and in the afternoon there was another session with the other half of the department. I stayed at the venue to work further on my design. I had a room where the Event department was also working".

Figure 4.10

Examples of different types of field notes

names, acronyms, project titles, facts, and any running commentaries, but also write notes referring to participants' body language, moods and what may have triggered a mood change. These notes help to identify new questions for the next session, better understand what the participant is doing or saying, go back to earlier topics and facilitate later analysis, by indicating interesting parts or quotations.

A way to determine whether something is important to include is to ask yourself whether that piece of information helped you in any way better understand what a participant said, their context, or what went on. Write down what participants say and do in the same order as you hear or see it.

How to plan and design the study | 77

What you feel. Also write down notes about your own feelings, emotions, mood, thoughts and reactions to what you are observing and hearing in each session. Remember, one of the goals of field research is for you to experience what your participants feel and do. These notes will help you empathize with your participants.

Keep memos or a personal diary documenting how you are experiencing the data collection part of the study. This is useful during data analysis to identify any personal assumptions or bias that you may have brought into the study.

What you think. Equally important are notes about your initial interpretations, insights, ideas, and inferences you draw based on contradictions between what you saw and heard, or any contradictory things you observed or heard at different times. Any inferred meaning from observations help identify personal assumptions and don't replace data analysis, but can later inform data analysis and be discussed within the team.

Make sure to distinguish interpretations from descriptive field notes either by changing colours or including the former between brackets.

Most of the notes will be handwritten notes, but some of them will also be visual or digital as you can also audio- and video-record each session, take photos, or draw sketches of what you see in the field. In any of these cases, test the equipment and the quality of recorded audio and image files beforehand. However, even if you are audio- or video-recording a session, you must take notes. Recording isn't a substitute for taking notes during or after a field session; it just removes the pressure of having to note every word a participant says or every detail you see.

Other types of data that you can capture to gain an understanding of people's lifestyle are photos from their workplaces, their desks, homes or neighbourhoods, a fridge inventory, outfits, wardrobe, and any other aspect of their lives that you think communicates something about them, their behaviours, or personality. The resulting types of field data often depend on the data collection method.

After each field session

Check the quality of the audio file just recorded, review any photos taken and your notes, to ensure that they make sense, and identify any immediate areas that you would like to expand or dig deeper into. Also check the quality of participants' responses and whether the data is useful. Table 4.2 provides examples of vague and high-quality field notes. Identify any changes to locations or questions that you think would be beneficial or any ambiguous

	VAGUE FIELD NOTES	DETAILED FIELD NOTES
Contextual interview notes written by design researcher	"Yes. Well, I don't know". She doesn't know the answer.	"Yes, well, I think with task two it was slightly easier, because it was my field, and I knew the associations that are well-respected. But with medicine I don't know… I don't know the important bodies of, you know, the important medical organisations, so, I don't know for sure which ones to look at".
Observational studies notes written by design researcher	The mall is crowded and cold.	Sunday 4:30 pm I'm standing next to computer store from where I can clearly see the information map. The mall is crowded, most people are families with young children (3–7 years old). Families seem relaxed, some stop to look at the "information map", others already seem to know where they are going.
Design probes notes written by participant	I used the tool. It was good.	Situation: 10am, at desk reading government documentation Thoughts: needing a change of pace, flipping thru cards is somehow intrinsically satisfying In addition to prompting for specific issues, the cards are useful as an aid to stepping-back from the details, to re-assess the big picture. A few times, while not intending to use the cards, I was doing some other task and vaguely reminded of one of the cards. So I quickly flipped thru the deck to find it, and the numerically adjacent cards allowed me to branch out to related tasks.

Table 4.2 Examples of vague and detailed field notes. When participants' responses are paraphrased or don't have much level of detail (second column), it can be really hard to identify needs during data analysis. Capture participants' exact words and include details (third column), this will help you identify needs later in the process.

wording that needs clarifying or rephrasing before the next session. Reflect on and write down any interesting non-verbal data you remember, for example, particular expressions or hesitations at specific questions. Later, these details help with the analysis and interpretation of the data and contextualize moods, actions, and opinions to specific situations, tasks, or activities.

Read your notes at the end of each day to make sure you understand everything you wrote (handwriting can be hard to read!), identify any unclear passages, and outline a few questions to clarify any parts of the story and stay on track with the purpose of the study.

Checklist

This is a list that summarizes key tasks in the research process, discussed throughout the chapter. Before going into the field, make sure that you and your team know the latest status of each task:

STATUS COMPLETE	STATUS PENDING	TASK
☐	☐	Determine reason for field study
☐	☐	Define goal (research question)
☐	☐	Create a picture reflecting initial understanding (conceptual framework)
☐	☐	Define the focus (unit of analysis)
☐	☐	Define who the participants are, and recruit them based on defined eligibility criteria using appropriate strategy
☐	☐	Identify the type of knowledge and insights needed, and select appropriate method/s
☐	☐	Assemble the team
☐	☐	Define study roadmap, put together study toolkit, find study workspace
☐	☐	Determine what aspects of the study need to be tested, and allocate time to conduct pilot studies

The next two chapters discuss data collection methods for exploratory and evaluation studies. Each method is presented in this format:

- **Purpose:** Brief description, outlining its goal and recommended uses for field research in information design.
- **Preparation:** The most relevant aspects for the effective use of each method are discussed. Sub-categories within this section indicate tips and specific materials or instruments that could facilitate its use.
- **Participants:** Direction and tips are provided for recruiting or interacting with participants in the field.
- **Data collection:** Basic steps to be followed, when using each method in the field to gather rich qualitative data, are provided.

From each method: Take only what is useful for your field study, goals and constraints. Focus on what you can do, while retaining as much rigour and richness as possible.

5 Gathering data: methods for exploratory studies

The following qualitative methods, summarized in Table 5.1, can be used alone or in combination to study people in their contexts. Each method is tailored to be suitable for the specifics of information design types of challenges and practitioners' needs.

General considerations:

- The more you can tailor a method and tasks (if any) to your participants' preferences, the more engaged they will be, and the richer the data you will gather.

- Each study is different; throughout it, issues will arise because time and resources are finite, and unexpected things occur.

- Use your creativity and trust your common sense to make the most of each situation.

- You can modify the direction of a study and identify new leads, based on whatever you have already collected during the first sessions.

- You can broaden the area of study, combine what you already collected with existing cases, or narrow the study by focusing on a theme or area for which you gathered the most data[1].

Observational studies

Purpose

The goal of an observational study is to directly observe and document people's actual behaviour. These studies involve objectively watching how people behave and perform actions, as well as noticing the dynamics and interactions in a place and what happens in a situation or phenomenon. Observational studies are the method of choice if you need to learn things about people that they would be unwilling to talk about or wouldn't consciously recall when asked directly, such as in an interview. Based on your degree of par-

---

1 Miles, M.B., Huberman, A.M., & Saldaña, J. (2013) *Qualitative Data Analysis*, SAGE.

METHOD	POSSIBLE USES IN INFORMATION DESIGN	DURATION
Non-participant observations	• To better understand a situation or how people do things • To learn how people use or interact with a new design in their environments [See Chapter 6]	• Ranges from one hour per day for various days (e.g. 5 days or every Friday for three weeks) to many hours (e.g. 6 hours) only one day
Shadowing	• To learn how people make decisions, what their mental models are, how they behave in social or work situations, and obtain a detailed log of actions • To learn how people use or interact with a new design in their environments [See Chapter 6]	
Participant observations	• To experience first-hand what people do, and record behaviour • To understand roles or perspectives, interactions, and reality from people's point of view • To learn how people use or interact with a new design in their environments [See Chapter 6]	
Contextual interviews	• To understand people's reflections on their own actions and experiences, intents and motivations, attitudes, values, perceptions, and feelings • To investigate infrequent actions or events that are hard to observe	• 45–90 minutes per session
Contextual inquiry	• To learn how and why a person is performing a specific task using a tool or product • To identify behaviours and sequence of actions in the use of a tool or product • To assess the use of a new solution in people's environments [See Chapter 6]	• 60–90 minutes per session
Diary studies	• To gather insights unreachable using the methods above • To understand people's needs and habits when they live in a different country or location • To capture people's interactions with a specific tool or product without interfering while data is gathered • To evaluate designs against requirements and in their intended context of use [See Chapter 6]	• Ranges from a few days (e.g. 5 days) to weeks (e.g. 2 weeks)
Design probes	• To better understand people's lifestyle, not just daily interactions with a specific tool or product • To capture people's general attitudes and behaviours when they live in a different country or location • To gather supplementary data to support idea development • To evaluate designs against requirements and in their intended context of use [See Chapter 6]	• Ranges from a few days (e.g. 5 days) to weeks (e.g. 2 weeks)
Collaborative workshops	• To come up with new ideas, or redesign something • To better understand how people think about a given problem, discipline or technology • To confirm whether what people say they do and what they actually do is the same • To include people as part of the information design process	• 90–120 minutes per session

Table 5.1 Overview of methods discussed in this chapter.

SAMPLE SIZE	LIMITATIONS	TYPE OF DATA GATHERED
• As many people as you observe related to the goal of the study (e.g. 30 in public place or 10 in a hospital hall)	• High level of unpredictability • Short preparation time but more time is needed to collect data • Researcher's bias	• Detailed descriptions of activities, situations, participants' behaviours and actions, and human interactions • Sketches, diagrams, photographs • Audio transcriptions, video recordings
• 2–5 people	• Find participants willing to be shadowed (e.g. the more senior a participant, the harder to coordinate visits to their office) • Subtantial coordination with participants is needed	
• Small samples (e.g. 2–5 people) or until saturation is achieved	• High level of unpredictability • Short preparation time but more time is needed to collect data • Researcher's bias	
• 5–12 people	• Some skill is needed to keep control of conversation • Participants might only say what you want to hear	• Quotations from participants about their experiences, feelings, opinions, frustrations, motivations, and goals • Sketches, diagrams, photographs • Audio transcriptions, video recordings
• 3–10 people	• Choose right tasks to examine • Capture everything participants say while observing what they do	• Quotations, and detailed descriptions of activities, situations, participants' behaviours, and actions • Sketches, diagrams, artefacts, photographs • Audio transcriptions, video recordings
• Minimum of 3 people; for best results 5 or more people	• Many components to design and coordinate • Lack of full control on data gathering • Need to keep participants engaged and committed	• Detailed descriptions of activities, situations, participants' behaviours and actions • Extracts from documents or entries • Sketches, diagrams, artefacts, drawings, collages, mock-ups, photographs • Audio transcriptions
• 4–10 people per session	• Design of engaging and useful tasks • Hard to coordinate with people with busy schedules. • Not everyone likes hands-on activities	

Gathering data: methods for exploratory studies

ticipation in the observed situation, the study can take the form of *non-participant observation*, *shadowing* or *participant observation* (Figure 5.1).

Non-participant observation is a form of "undercover" (covert) observation: you remain outside, not interfering in the situation you are observing; you are a spectator, and participants don't know you are observing them. This type of observation is useful for learning how people do things, so you can create designs that better support their process and task flows (e.g. withdraw money from an ATM, know where to go in an airport). When you conduct this type of observation sit aside to be unobtrusive without making people aware, for example when observing people in public places, watching people moving through a museum, or observing interactions between patients and nurses in a hospital.

Shadowing is a form of overt and structured observation, in which the participant knows that you are observing them but you don't intervene in the situation being observed or take an active role[2]. Rather than only relying on what you see and hear, you become the participant's shadow and follow them over an extended period to learn what they actually do in their everyday lives[3]. The level of planning should be minimal as the participant should "lead the way". You can ask unplanned questions to fully comprehend each situation (e.g. after a phone conversation, what did the other person say?) and help participants articulate tacit knowledge and skills (e.g. ask them to think aloud while performing a task). There are many ways you can spend time with a participant such as commuting together to work (by train, walking), sharing meals, sitting with them in the office, or using the gym together. Shadowing is useful for gaining better understanding of how people make decisions, their mental models, and how they behave in social situations. These insights can help design products and services that fit better into people's routines. This method is particularly useful when designing something for people of, or living in, a different culture, and you can afford travelling to where they are to conduct the study.

Participant observation is a form of *overt* or *covert* observation, in which you become a 'player'[4] in the situation being observed, by participating in any activities that occur. When it is *covert*, it is often referred to as "walk a mile on your customer's shoes". In either case, when using this method, you take the role that you are studying. For example, you could become a patient

---

2 Sommer, B. (2006) *Participant observation*, Department of Psychology, University of California [online], Available at: http://psc.dss.ucdavis.edu/sommerb/sommerdemo/observation/partic.htm [Accessed 13 January 2018].
3 McDonald, S. (2005) Studying actions in context: A qualitative shadowing method for organizational research, *Qualitative Research*, 5(4), 455–473.
4 Sommer, B. (2006) *Participant observation*, Department of Psychology, University of California [online], Available at: http://psc.dss.ucdavis.edu/sommerb/sommerdemo/observation/partic.htm [Accessed 13 January 2018].

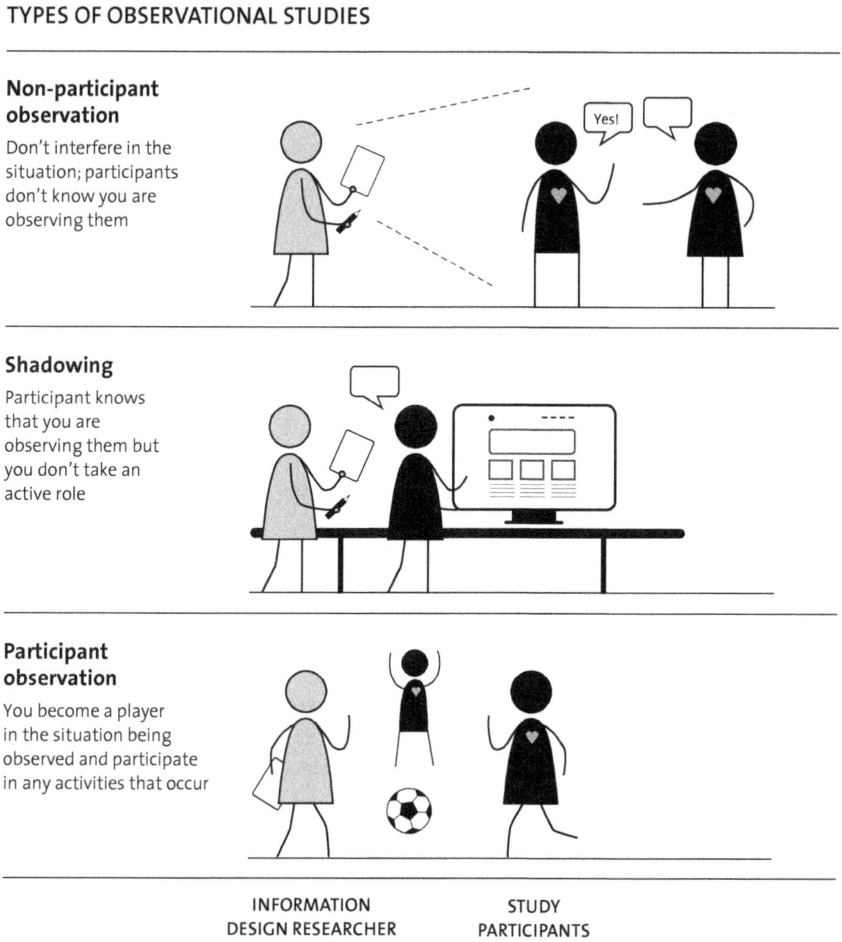

TYPES OF OBSERVATIONAL STUDIES

Non-participant observation
Don't interfere in the situation; participants don't know you are observing them

Shadowing
Participant knows that you are observing them but you don't take an active role

Participant observation
You become a player in the situation being observed and participate in any activities that occur

INFORMATION DESIGN RESEARCHER   STUDY PARTICIPANTS

Figure 5.1
Overview of the three types of observational studies discussed in the chapter.

in a hospital, trying to understand what nurses are saying, or a commuter trying to navigate the transport system. Another use of this method is to understand people's interactions. For example, if you are already part of the group, you can gain better understanding of the internal dynamics of a company, become familiar with people's vocabulary and terminologies, understand businesses' interaction with customers, or discover common gaps in work chains.

Conducting "undercover" observational studies could raise ethical issues, as you would need to be very careful how you use data and insights, but these three types of observational studies are used in information design, often conducted in combination with *contextual interviews*. For examples of how this method has been used in information design see Case Study 1 (participant observation), and Case Studies 2 and 4 (non-participant observation) in PART IV.

Preparation

There is no right way to conduct an observational study, but it does require some practice[5] and making decisions about specific dimensions representing variations that can occur in the field. These dimensions are as follows:

1. The goal: What do you want to get out of the observation? Clearly define the purpose and determine the specific goal of the study.

2. The focus: What will you observe and document? Depending on the goal, observations can have a narrow focus, only targeting the understanding of a single element or small parts of what is happening, or a broad focus, involving a holistic view and understanding of a situation.

3. Your role: What will be your role as an observer? Will participants be aware that you are observing them? Although it is better to determine your role in advance, your degree of participation can change throughout the study.

4. The location: Where will the observation take place? You can conduct observations in people's homes or workplaces, or in open public places (e.g. a park), hospitals, shopping centres, etc. The location depends on the type of observational study you conduct. In non-participant observation, you must find a place with good visibility and determine where will you put yourself while observing, whether you will change places or stay in the same place. It is important to visit the location you will be observing beforehand to find the most suitable place and to determine any factors that may affect the study (e.g. time slots, activities).

5. The duration: How long will the study last? Observation lengths depend on the time and resources available but should last long enough to address the goal of the study (e.g. observe how people complete a transaction or observe how people find their way out from a museum). You may need to observe for hours over one day or for shorter periods over many days. In the case of shad-

---

5 Patton, M.Q. (2002) *Qualitative Research & Evaluation Methods*, 3rd edn, Thousand Oaks, CA: SAGE; Blandford, A., Furniss, D., & Makri, S. (2016) Qualitative HCI research: Going behind the scenes, *Synthesis Lectures on Human-Centered Informatics*, 9(1), 1–115; Flick, U. (2009) *An Introduction to Qualitative Research*, 4th edn, Thousand Oaks, CA: SAGE.

owing, the duration also depends on participants' availability. If you need to observe them performing a specific activity, the study should preferably last until the activity is completed or you gather enough information to understand how they perform the activity.

Participants

Unless you are conducting non-participant observations, the sample size is relatively small (two to five people). In the case of shadowing, you need to clearly identify who your potential participants are (so avoid using convenience sampling). When conducting overt observational studies, seek informed consent from participants and, if visiting workplaces, permission from the organization, as you may have to attend meetings or places only open to authorized personnel. If informed consent isn't possible, you shouldn't use people's data in any future reports or presentations. When shadowing, reach out to participants early on and coordinate where and when would be the best places, times, and days to observe; contact them in advance to explain the study and what it involves. Offer a face-to-face meeting and start building a relationship and rapport. Spend one or two days getting to know them and what they do first; otherwise, it will be harder to fully understand everything your participants say, and take meaningful notes.

Data collection

Observing in the field can be overwhelming because there are so many things to watch at the same time. To gather high-quality data, you must move away from ordinary *looking* and closer to systematic *seeing*. The difference between these two actions is that, when looking, you focus on collecting raw data that is in front of you, while systematic seeing is a more rigorous way of looking. It involves observing a situation in layers and progressively narrowing the focus from general to specific, selecting what is important, and then identifying patterns[6]. One way to develop systematic seeing is conducting observations of various depths[7]:

- **Descriptive observations:** First, get a sense of the general picture. Observe the setting and gain an understanding of the context (where); pay attention to the (planned and unplanned) activities taking place in that setting (what, when, how), the people who participate in those activities, and their experiences and behaviours (who).

---

6 Roam, D. (2013) *The Back of the Napkin: Solving Problems and Selling Ideas with Pictures*, London: Portfolio.
7 Flick, U. (2009) *An Introduction to Qualitative Research*, 4th edn, Thousand Oaks, CA: SAGE.

	CHARACTERISTICS	DETAIL
Space (e.g. office, meeting room, living room, museum, etc.)	Colours	What colours are used in the space?
	Textures	What is the texture of the chairs?
	Smells	What is the predominant smell?
	Sizes	Are people wearing the right t-shirt sizes?
	Dimensions	Is the space big enough for that activity?
	Sounds	Is there ambient music?
	Objects	Are they new, modern, old?
People	Physical characteristics	Height, weight, hair and eyes colour
	Demographics	Gender, age, race
	Social interactions	How are the interactions? Interactions with one person, with many people, talking only, talking and laughing, etc.
	Physical appearance	Outfits, neatness, hairstyle
	Body language	Arms crossed, arms moving, facial expressions
	What they are doing	Talking, sitting, standing
	Feelings and attitudes express through their bodies	Yawning, standing still, fidgeting

Table 5.2 Aspects you should pay attention to when conducting an observational study. Based on Lawrence, N.W. (2013) *Social Research Methods: Qualitative and Quantitative Approaches*, 7th edn, Harlow: Pearson New International Edition.

- **Focused observations:** Gradually shift the focus to a few specific objects, interactions, problems, or issues. To do this, narrow the focus of what you are seeing to only those aspects of the context that are key for your study or that you find may be relevant.
- **Selective observations:** Lastly, selectively focus your attention onto things and dimensions to help you answer specific questions and themes. For example, pay attention to and look only at those specific aspects that caught your attention and/or are directly related to the goal of the study. You could also observe situations that are extremely different.

To fully address these three types of observations, spend time paying attention to each *space* of the environment, the *people* and their interactions, and capture their *characteristics* and every *detail* (Table 5.2). At first, when you start observing, everything may appear relevant. So, until you get used to systematic seeing, practise this sequence of observations in a very structured way. Creating *an observational guide* will help you have some structure in the field.

Figure 5.2 provides a visual summary of how to plan and conduct an observational study.

HOW TO PLAN AND CONDUCT AN OBSERVATIONAL STUDY

① PLAN

☐ **DEFINE GOAL**
What do you want to get out of the observation?

☐ **DEFINE FOCUS**
What will you observe and document?

☐ **DEFINE YOUR ROLE**
What will be your role as an observer?
- Non-participant observation
- Shadowing
- Participant observation

☐ **DEFINE LOCATION**
Where will the observation take place?

☐ **DEFINE DURATION**
How long will the study last?

② PREPARE

Create observation guide

Create note templates

Bring notebook

Bring coloured pens, pencils, and highlighters

Bring camera (e.g. smartphone camera)

Recruit and coordinate with participants (if applicable)

③ TEST BEFORE GOING INTO THE FIELD

☐ **PILOT STUDY**
How can you ensure that you will collect the data you need?
- If you encounter problems, revise ① or ②
- Otherwise, start study in the field

④ GO INTO THE FIELD

☐ Conduct descriptive observations

☐ Conduct focused observations

☐ Conduct selective observations

Figure 5.2 Key steps to plan an observational study.

Contextual interviews

Purpose

The goal of contextual interviews is to gain a deep understanding of people's reflections on their own actions and experiences, intents and motivations, attitudes, values, perceptions, and feelings. In other words, to elicit

Gathering data: methods for exploratory studies | 89

rich descriptions and possible explanations of how they make sense of the world and experience particular events. Unlike surveys, contextual interviews are semi-structured and all questions aren't pre-determined or the same for each participant. Rather, contextual interviews are more similar to a structured conversation, with a particular focus and purpose, but having an open-ended approach[8,9]. As you will be in the participant's environment, the interview may be interrupted or take an unexpected turn for which pre-planned questions won't be as helpful as asking spontaneous questions to learn more about the new situation (Figure 5.3).

Contextual interviews can also be used to learn about the specifics of a project topic by interviewing *key informants*[10], rather than the intended audience. Key informants are seasoned professionals, leaders, or experts in the topic of the design project or an aspect of the design idea. Their insights can provide better understanding of the topic and identify areas to explore further with the intended audience. For examples of how this method has been used in information design see Case Studies 1, 2, 3, and 4. Case Study 5 presents an example of contextual interviews conducted with key informants.

Preparation

To prepare a contextual interview:

1. Determine focus and direction. What key areas will you explore to holistically understand the given situation? One way to do this is to identify the main topic of the situation, and then break it down into more manageable sub-topics or themes. List the aspects of the situation that you already know a lot about and highlight those about which you know very little. Identify your assumptions and gaps in your understanding. Create a hierarchy, specifying what sub-topics are more important for the design project, and determine a flow; that is the order in which you will explore each sub-topic.

2. Generate questions. What type of questions will you ask to investigate each identified theme or sub-topic? For each theme or sub-topic, you can formulate questions about behaviours, experiences, opinions, feelings, and general knowledge. Also generate questions to capture participants' demographic

---

8 Blandford, A., Furniss, D., & Makri, S. (2016) Qualitative HCI research: Going behind the scenes, *Synthesis Lectures on Human-Centered Informatics*, 9(1), 1–115.
9 Beabes, M. & Flanders, A. (1995) Experiences with using contextual inquiry to design information, *Technical Communication*, third quarter, 409–420.
10 Marshall, M.N. (1996) The key informant technique, *Family Practice International Journal*, 13(1), 92–97.

CONTEXTUAL INTERVIEW SETUP

Figure 5.3
When conducting contextual interviews, you go to the participant's home or workplace, and sit on the side rather than in front of the participants to avoid intimidation.
Be prepared for interruptions and the unexpected; questions may change from one participant to the other.

INFORMATION DESIGN RESEARCHER   STUDY PARTICIPANT

information[11]. The type of questions and how you ask them are also key aspects for closer examination:

- **Type of questions.** Regardless of the topic, questions should be *open-ended*. They increase the depth of responses and help gather richer data than closed-ended questions because they allow participants to respond in any direction and using any words they want. You can use the basic questions—"who", "what", "where", "when", "why", and "how"—to ask about different dimensions of a situation or topic and obtain a more complete picture. Avoid asking *leading questions* (e.g. Isn't it right that you found the form hard to complete?), *binary questions* that can be answered with one word (e.g. Do you struggle with the old interface?), or those that constrain participants' way of describing their experience by providing pre-determined answers (e.g. Are you feeling happy or sad?).

- **Wording.** Using appropriate wording makes the difference between a clear and a confusing question. Avoid unnecessarily complex words. Instead, become familiar with the language and jargon your participants use, and employ that same language when wording interview questions. This is extremely important when interviewing people of a different culture. If participants don't understand the questions, they may feel uncomfortable or ignorant, as they may not ask for clarification. For example, if you are interviewing American participants about their cooking habits, use the word "stove", rather than "hob".

[11] Patton, M.Q. (2002) *Qualitative Research & Evaluation Methods*, 3rd edn, Thousand Oaks, CA: SAGE.

Gathering data: methods for exploratory studies | 91

3. Create interview guide. Determine a list of topics and sample questions to explore during the interview. You can also specify the order or flow, in which you would like to approach each topic, even if this may vary from one participant to another.

4. Pilot study. Test the interview guide before the actual study to check clarity, consistency, and even the flow of questions. Going into the field, having careful planned and practised what to ask, helps you to be more relaxed and confident, and to keep control during the interview.

Participants

The sample size depends on how much time you will spend with each participant, and whether you will conduct multiple sessions with the same participants, or just one session with many different participants. Sometimes, the former is preferred as you get to know the participants in more depth. In either case, sit on the side rather than in front of the participants to avoid intimidation. Sometimes finding a quiet space to conduct the interview may be hard, as for example, if you are interviewing a participant in their home or workplace there may be other family members interrupting the conversation or phones ringing. If interruptions occur, be prepared to get the interview back on track and assess their impact on the participant's responses.

Data collection

A basic interview structure has the following parts:

1. Introduction. State the aim of the interview and explain what you will do with the collected data. Give just enough information about the study but not too many details that could bias participants' responses or make them think you want to hear a specific view on the topic. Also, reassure participants that your goal is to understand; you aren't evaluating or judging their responses.

2. Warm-up questions. Start by asking one or two questions, not necessarily related to the study topic, to initiate conversation and make the participant feel relaxed and comfortable. For example, you can ask about their background or to have a tour around their workspace or home. Ask whether they are happy for you to record the interview and take a few photographs of the environment (e.g. desk, technology use, office). Then, you can ask demographic questions, but you can also collect this information via a short questionnaire at the very beginning of the interview or when recruiting participants.

3. Core in-depth questions. Ask one or two general questions about the study, and then continue by asking more specific topic-related questions.

- **Listen.** Pay attention to what participants say and try to ensure the next question builds on what they just said. If thinking about your next question while participants are still talking, you aren't actually listening to what they are saying. Notice the terminology and words they use to answer the questions and try to use that same language throughout the rest of the interview.

- **Ask why.** Be genuinely curious. Even when you think you know the answer, ask people why they do or say things. A conversation that started from one question should continue for as long as it needs to. Ask for specific details, elaboration, or clarification, but ask one question at a time.

- **Be specific.** Avoid saying "usually" or "often" when asking a question; instead, ask about a specific instance or occurrence, such as "tell me about the last time you..." or "describe what you did yesterday when you arrived to the office".

- **Be neutral.** Avoid showing very positive or negative expressions, or sharing personal bias to what participants say. Use neutral language like "Go on..." or "Aha...", or nod your head to encourage participants to express themselves and create a non-judgemental space.

- **Encourage stories.** Request stories, best and worst experiences, concrete examples or to draw or show you the item (e.g. object, space) they are talking about. This helps participants recall their memories and experiences (that is, their explicit knowledge). Whether or not the stories people tell are true, they reveal how they think about the world, and this is the kind of data you need to inform the design process.

4. Closure. Wrap up, thank the participant and explain any follow-up steps (e.g. if you will contact them again with further questions or follow-up session).

One goal of this type of interview is to gain such an understanding of your participants that you can empathize with them. Pay attention to non-verbal cues, and be aware of body language and emotions. Also, don't be afraid of silence and avoid suggesting answers to your questions when participants pause. This can unintentionally make them say things to agree with your expectations. During a pause, allow silence, so the participant can reflect on what they have just said; they may reveal something deeper. Just wait until they start talking again.

Figure 5.4 provides a visual summary of how to plan and conduct a contextual interview.

HOW TO PLAN AND CONDUCT A CONTEXTUAL INTERVIEW

1. PLAN

- **DEFINE GOAL**
  What do you want to get out of the interview?
- **DEFINE FOCUS & DIRECTION**
  What key areas will you explore?
- **GENERATE QUESTIONS**
  What type of questions will you ask?
  Are the questions open-ended?
  Is the wording appropriate?

2. PREPARE

- **Create** interview guide
- **Create** note templates
- **Bring** notebook
- **Bring** coloured pens, pencils, and highlighters
- **Bring** camera (e.g. smartphone camera)
- **Recruit and coordinate** with participants

3. TEST BEFORE GOING INTO THE FIELD

- **PILOT STUDY**
  How can you ensure that you will collect the data you need?
  - If you encounter problems, revise 1 or 2
  - Otherwise, start study in the field

4. GO INTO THE FIELD

- Introduce study and goal
- Ask warm-up questions
- Ask core in-depth questions
- Close study

Figure 5.4
Key steps to plan a contextual interview.

Contextual inquiry

Purpose

The goal of contextual inquiry is to better understand how and why a person is performing a specific task[12] in their environment (e.g. office, studio, house), using a specific tool or product, and its influence in their lives. To elicit this understanding, this technique combines interviews and observations that have a clear focus and direction (Figure 5.5). You ask the participant to verbalize what they are thinking and doing; this allows you to concurrently investigate their thinking, doing, and reasoning[13] which reveals tacit thoughts and latent needs. Data is often gathered over many field visits.

When shadowing, you observe participants without interrupting what they are doing, only asking questions for clarification. In contextual inquiry, when a participant isn't describing in detail the task they are performing, you encourage them to verbalize their thoughts and explicitly articulate what they are doing, while still performing the given task. After the task, you can also ask further questions to better understand any unclear passages of their description.

Typical uses of contextual inquiry in information design are to learn people's impressions of the use of an existing design, learn specific details of how people do their jobs, identify new ways of performing a task, or understand problems with a new design. For example, observe and learn people's experiences when completing tax forms to identify what isn't working for them and why.

Preparation

To prepare a contextual inquiry[14], you follow similar steps to those for observational studies, in that you must identify the location where the study will take place and determine the number of field visits to make. In this case, your interview guide will have even less structure than when conducting just contextual interviews. Instead, the *focus* structures and gives direction to the conversation. It is important to determine the specific activities and tasks you will observe and discuss with each participant.

---

12 ibid.
13 Ericsson, K.A. & Simon, H.A. (1993) *Protocol Analysis,* Cambridge, MA: MIT Press.
14 Beyer, H. & Holtzblatt, K. (1998) *Contextual Design: Defining Customer-Centered Systems,* San Francisco: Elsevier.

CONTEXTUAL INQUIRY SETUP

Figure 5.5
When conducting a contextual inquiry, you identify needs while the participant performs a specific task in their home or workplace.

INFORMATION DESIGN RESEARCHER   STUDY PARTICIPANT

Participants

The sample needed is also small, but you will spend long periods with each participant. Depending on the size of the project and complexity of the topic, you may need two or six, but more than ten participants is rare. When recruiting participants, clearly explain the structure of the study and the time commitment required on their behalf. Ensure you schedule visits in advance. Depending on the task or activity you will observe, you may need to schedule more than one visit. Emphasize that seeing them on a typical day and performing the given activity in their regular environment is part of the study.

Data collection

The structure of a contextual inquiry session involves:

1. Opening interview. Introduce the goal of study, ask a few questions to build rapport, and then request permission to record the session and take photographs.

2. Transition. Explicitly explain that the study will now move from a traditional interview to a master-apprentice session. You will adopt the role of novice or apprentice, wishing to learn from the participant, who is the expert or master (on the task and domain). Also, stress that your role will be mostly observing them perform a task, but that you may occasionally interrupt them. Clarify when would be best to do so, without disturbing their work.

3. Ongoing observation and discussion. During this part of the session, the focus should be on the task the participant is performing (master) and you observing (apprentice). Sit and watch, and ask questions to know why they are doing those activities and in that particular order or for clarification when you don't understand what they are doing, even if you think that your questions are irrelevant. These are three principles you can follow:

- **Use artefacts:** Bring artefacts—anything they describe, create, or work with—into the conversation to trigger memories and anchor the conversation with concrete examples rather than abstractions.

- **Observe context:** Pay attention to other people involved in the scene, whether they contribute to the task or intervene at any point in the process of completing the task. Discuss the experience and the complexity of the tasks, to describe the easiest and hardest moments, and to explain why they felt that way.

- **Create partnership:** In contrast to contextual interviews, where you should remain in control, now participants lead the conversation by explaining what they are doing. Your role is to collaborate with them and ask questions about what you are observing, to understand their motivations, struggles, and strategies to succeed.

4. Summary. Share with participants your assumptions and interpretations of what you think they did and the meaning of their words and actions. Participants' comments and corrections to your interpretations are key to constructing an accurate understanding of their actions and how these contribute to their life. This will help determine what they need.

Figure 5.6 provides a visual summary of how to plan and conduct a contextual inquiry study.

Design probes and diary studies

Purpose

The goal of design probes and diary studies is to give participants the active role in recording their experiences, activities, feelings, behaviours, and thoughts during a long period of time, while you minimize your influence on them and remain an 'outsider'[15]. Participants can gather data in specific

---

15 Gaver, B., Dunne, T., & Pacenti, E. (1999) Design: Cultural probes, *Interactions*, 6(1), 21–29; Mattelmäki, T. (2008) *Design Probes*, 2nd edn, Vaajakoski: University of Art and Design Helsinki.

HOW TO PLAN AND CONDUCT A CONTEXTUAL INQUIRY

1. PLAN

- **DEFINE GOAL**
  What do you want to get out of the contextual inquiry?
- **DEFINE FOCUS & DIRECTION**
  What key areas will you explore?
- **DESIGN TASKS & ACTIVITIES**
  What specific tasks will you observe?

2. PREPARE

- **Identify** tasks
- **Create** note templates
- **Bring** notebook
- **Bring** coloured pens, pencils, and highlighters
- **Bring** camera (e.g. smartphone camera)
- **Recruit and coordinate** with participants

3. TEST BEFORE GOING INTO THE FIELD

- **PILOT STUDY**
  How can you ensure that you will collect the data you need?
  - If you encounter problems, revise 1 or 2
  - Otherwise, start study in the field

4. GO INTO THE FIELD

- Start with an interview
- Introduce study and move to master-apprentice session
- Observe and discuss
- Summarize and share understanding

Figure 5.6
Key steps to plan a contextual inquiry study

pre-determined ways (e.g. taking photos, writing entries, drawing) and time intervals, such as their own time, at specific times of the day (e.g. morning, midday, evening), or in response to a reminder (e.g. received by text message or email) (Figure 5.7). These methods focus on studying habits, attitudes and motivations, usage scenarios, changes in behaviours, people's journeys, and processes to achieve a goal.

Typically, in information design, design probes and diary studies can be used to gain a more accurate picture of people's needs and behaviours (e.g. in the workplace, in the gym), to understand people's interactions with a product or service (e.g. use of a device, such as TV remote control), to better understand how people complete specific (e.g. cook dinner during weekdays) or general activities (e.g. use social media, use public transport), to examine people in different locations, or to understand an intermittent phenomenon (e.g. prepare for a trip). For example, if you design a diary study to better understand the use of phone apps in college students, participants will document each time they use an app and write an entry explaining why they use it and for how long. Resulting insights can be particularly useful to better understand their behaviours and identify key features a future app should have supporting ideation.

Preparation

These studies can take many forms, based on the goal, resources, and time frame available, but all include a package with a set of data collection instruments and tools, given to participants to encourage engagement and responses[16]. Figure 5.8 provides an example of a design probe package with data collection instruments. Planning a self-documentation study isn't hard, but it requires careful preparation because there are many components and moving parts that must be coordinated and tested. Based on the study's goals, first define a suitable structure for the study; that is, what you will ask participants to do and document, and for how long. Then, design a data collection package, including objects, tasks, and instructions. After data collection ends, prepare an interview guide with follow-up questions to learn specific details about what each participant reported.

1. Create package. A self-documentation package—traditionally a bag, folder, or envelope—contains a set of *objects and instruments*, and *tasks* for participants to document and collect personal data.

- Objects and instruments can be anything (e.g. postcards, illustrated cards, stickers, notebooks, diaries, disposable cameras, smartphones) that encourages inspiration and playfulness, and reminds participants to record written, verbal or visual data during the given time. These objects can also be designed in *digital format* (e.g. text messages, emails, online forms), minimizing participants' time to collect and log data and therefore increasing their commitment to the study, which used to be a

[16] Bolger, N., Davis, A. & Rafaeli, E. (2003) Diary methods: Capturing life as it is lived, *Annual Review of Psychology*, 54(1), 579–616.

DESIGN PROBES AND DIARY STUDIES STRUCTURE

Figure 5.7
When conducting design probes and diary studies, you minimize your influence on the participants as you remain an outsider, and they have the active role of gathering data. You will only meet with participants at the beginning (Step 1) and at the end of the study (Step 4).

frequent major obstacle for traditional self-documentation studies in professional contexts. Currently, you can design fully digital design probes and diary studies, involving the use of smartphones (e.g. text messages) or simply emails to send reminders, and the use of apps (e.g. WhatsApp), online platforms (e.g. specific sites, blogs, Twitter, Facebook), or file hosting services (e.g. Dropbox, Google Drive) to capture personal data.

- Tasks involve exercises to guide participants through the collection of their personal data by, for example, asking them to take photos of their

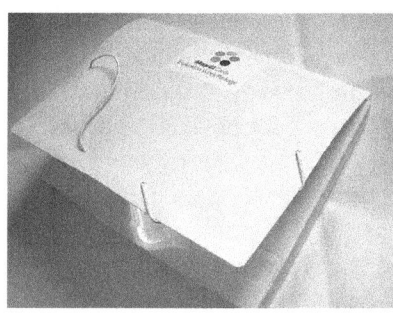

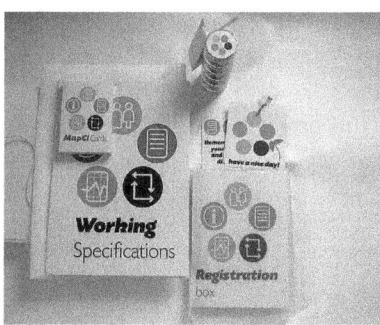

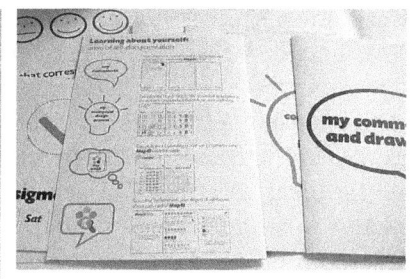

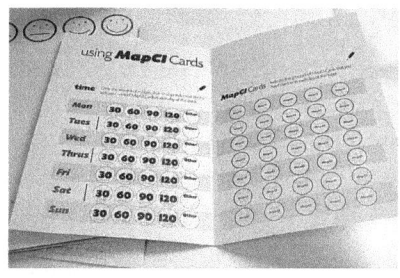

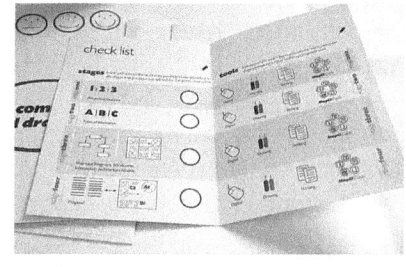

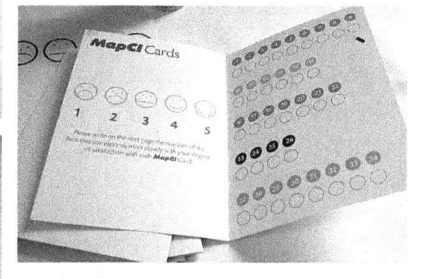

Figure 5.8

Package and set of data collection instruments for a design probe study. This study was conducted to better understand how designers work. Participants had to complete specific forms based on how and when they performed a task. In a second study, many of these objects were redesigned in digital format to facilitate participants' engagement in the study.

Gathering data: methods for exploratory studies | 101

environments, write diary entries about their activities, draw maps describing how they commute to work every day, or create collages to represent what they eat for breakfast during a week. Tasks can take the form of open-questions (e.g. what did you eat for breakfast?) or statements associated with the topic being studied (e.g. describe what you ate for breakfast); they can also constitute forms to complete, using stickers to represent their feelings when performing a particular task (e.g. when they wake up) or one-word pages for participants to create diagrams, indicating their emotions in specific situations.

While there are no rules for the creation of self-documentation packages, objects, and tasks, as exercises and visual design vary from project to project, the format and design should respond to both the purpose of the study and participants' needs and interests. The packages have a direct influence on participants' motivation and level of response. It is important to create them from the participants' perspective and, if relevant, to even customize each package. For example, if you are conducting a study to explore older populations' habits, they may feel more comfortable receiving a traditional package by regular mail than a digital one by email.

2. Create instructions. Create a simple script to give to participants at the beginning of the study to help them structure data-capture and elicit the necessary behavioural reflection. Be explicit about the type of information they need to capture and log, and provide examples that are related but broad enough so you don't bias participants.

3. Pilot study. Once you have a draft version of the package, piloting is essential to reveal lack of clarity in the tasks and instructions and identify obvious flaws prior to the full experience. One way to pilot your package may be to ask one or two people (with similar characteristics to those of the participants, if possible) to go through the logistics of the study for a shorter period. They should read the instructions and use the self-documentation package to complete the tasks. This helps identify unclear phrasing and verify whether each component meets the desired objectives.

Participants

Unlike other methods, recruiting participants can be challenging because these studies demand a large commitment of participants' time. While you need a small sample, it is important to carefully determine eligibility criteria and select participants who are reliable and highly motivated. For example, if your goal is to gain further understanding of a defined phenomenon (e.g. how young adults in Denmark use a digital device), key recruitment criteria may be country of residence, age, and ownership and regular use of the digital device. Once you have found potential participants, organize short online or face-to-face interviews to gauge their level of motivation and engagement with the study, but recruit more participants than you need, as it is common for a few of them to drop out. Some factors that increase participants' motivation are the study topic, people-centred packages, and compensation, even if minimal.

Data collection

In general, these studies consist of four parts:

1. Initial interview. This is an initial meeting to give each participant the package and explain how the study works, the time frame, and what it is expected of them. If possible, conduct a face-to-face interview, but if participants are in a different city, organize a remote meeting. In the latter case, if you designed a physical package, make sure to send it in enough time for them to receive it before this initial meeting. When briefing participants about the tasks, it is vital to stress the importance of documenting everything requested and at the given times. During this first encounter, also collect participants' demographics if you haven't already done so at recruitment.

2. Participants' reflective time. This is the time you have allocated for participants to self-collect data. To monitor the study, during this period you:

- **Send reminders.** Remotely send reminders to participants and/or further instructions, as agreed with each participant, for example, by text message, WhatsApp or email. While the success of this method relies on *rigorous* monitoring to ensure the gathering of the appropriate data, you must ensure you send sufficient messages but not too many to cause fatigue or annoyance. For example, avoid sending messages too early or too late in the day; lunchtime tends to be a good time, but it is best to agree with each participant beforehand when would be less disruptive for them to receive these messages. Be patient; don't intervene during this part of the study beyond the agreed reminders, unless participants contact you with questions or you haven't heard anything from them in more than 24 hours. Remember that, often, unplanned things occur, and participants can be late with their daily reporting.

- **Evaluate incoming data.** Review the data as it arrives and contact participants if they are collecting data which is completely unrelated to what has been requested. Sometimes, participants' behaviours can be the result of unclear prompts or questions; if this is the case, you can always stop the study and send them new tasks. This is preferable to proceeding to the end and having useless data

3. Revision of submitted data. Review the data sent by each participant and prepare questions to learn more about interesting or unclear parts.

4. Final interview. This is a final meeting with each participant at the end of the assigned period to debrief them about the experience, ask questions about the data, and collect the package. Also, ask participants about any stories that may have emerged during the study, and review the entries with them to unpack any cryptic passages.

Figure 5.9 provides a visual summary of how to plan and conduct a design probe or diary study.

Collaborative workshops

Purpose

The goal of *collaborative workshops* is to help participants discover unknown and undefined needs or identify specific steps in the way they perform a task that can then inform the information design process by generating ideas and opportunities[17]. Collaborative workshops are group meetings, often conducted in people's workplaces where participants work or play (Figure 5.10). Unlike observational studies and contextual interviews, in collaborative workshops, participants don't only express their ideas with words; they create some form of artefact to communicate their thoughts and dreams or engage in activities to generate and analyse ideas. After completing given activities, participants explain what they did and why. This process of thinking and creating helps elicit tacit knowledge, identify hidden dreams, and reveal latent needs[18].

In a collaborative workshop, your role is to introduce activities and facilitate a conversation to better understand the ideas behind what participants did and help them articulate their process. You won't have an interview guide or list of questions; but you should spend time with each participant to discuss the outcomes of the activities and better understand what is meaningful to them. While your main role is to lead and facilitate the meeting, you may occasionally contribute with your ideas and build

---

17 Sanders, E.B.N. (2002) From user-centered to participatory design approaches, Chapter 1 in Frascara, J. (ed.) *Design and the Social Sciences: Making Connections*, London: Routledge; Mattelmäki, T. (2008) *Design Probes*, 2nd edn, Vaajakoski: University of Art and Design Helsinki.
18 ibid.

HOW TO PLAN AND CONDUCT DESIGN PROBES OR A DIARY STUDY

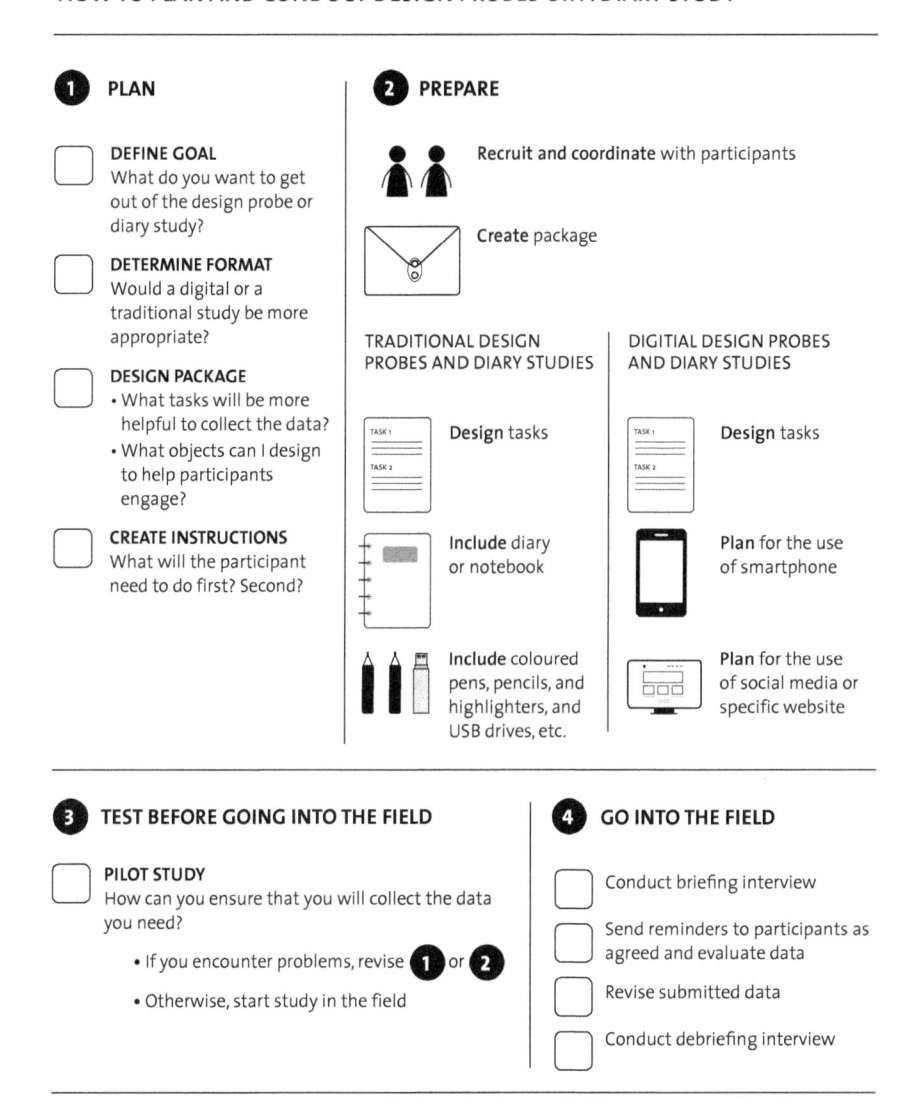

1 PLAN

DEFINE GOAL
What do you want to get out of the design probe or diary study?

DETERMINE FORMAT
Would a digital or a traditional study be more appropriate?

DESIGN PACKAGE
- What tasks will be more helpful to collect the data?
- What objects can I design to help participants engage?

CREATE INSTRUCTIONS
What will the participant need to do first? Second?

2 PREPARE

Recruit and coordinate with participants

Create package

TRADITIONAL DESIGN PROBES AND DIARY STUDIES

Design tasks

Include diary or notebook

Include coloured pens, pencils, and highlighters, and USB drives, etc.

DIGITIAL DESIGN PROBES AND DIARY STUDIES

Design tasks

Plan for the use of smartphone

Plan for the use of social media or specific website

3 TEST BEFORE GOING INTO THE FIELD

PILOT STUDY
How can you ensure that you will collect the data you need?
- If you encounter problems, revise 1 or 2
- Otherwise, start study in the field

4 GO INTO THE FIELD

Conduct briefing interview

Send reminders to participants as agreed and evaluate data

Revise submitted data

Conduct debriefing interview

Figure 5.9
Key steps to plan a design probe or diary study.

on others. For example, during a session, you will give participants tools to create mock-ups of "ideal" solutions to a problem at hand. "Ideal" solutions should be something that participants would love to use in a "perfect world" to improve their current situation. While participants will lead the creative process, you will ask them to explain why they are building their ideal solutions in that particular way. From observing participants' building process and listening to their explanations, you will learn what matters to them and other information that would not be elicited from only asking questions. A version of this method was used in Case Study 2 to help identify people's mental models.

Preparation

As with design probes and diary studies, planning collaborative workshops involves the creation of a toolkit, including the definition of activities, design of instruments, and selection of a set of tools for each session.

1. Define activities. Most activities used in collaborative workshops are *generative*[19]; that is, they encourage active participation and make the implicit more explicit. This helps participants become more aware of their everyday experiences and notice and articulate hidden feelings. These activities are based on the principle of letting participants think while making things (e.g. objects, drawings, collages) and then tell a story about what they made[20]. Activities should reflect the goal of the study, but exercises can take any form and vary from one study to another. For example, exercises can be developed to help participants express current or aspirational emotions (e.g. ask participants to describe an experience, using words that depict feelings), describe processes (e.g. ask participants to describe how they do or would like to do something, using action or activity words), or imagine ideals (e.g. ask participants to imagine a business or an activity in the future to better reflect their needs)[21].

Generative activities can be used for different purposes—articulate experiences and feelings, tell stories, encourage discussion, build understanding—in different steps of the research process. This section focuses on three activities you can use in a collaborative workshop, other similar activities are discussed in Chapter 8:

- **Collage activities.** Creating collages helps participants express themselves through images and words, and discover emotions, feelings, ideals, or dreams about the topic under exploration. Making collages is an abstract activity that helps participants articulate how they describe current and envision aspired experiences. For this activity, you need large quantities of magazines, coloured paper, and markers and can add

---

[19] Sanders, E.B.N. & Stappers, P.J. (2014) Probes, toolkits and prototypes: Three approaches to making in codesigning, *CoDesign*, 10(1), 5–14.
[20] ibid.
[21] Gage, N. (2012) *Making Emotional Connections Through Participatory Design* [online], Available at: http://boxesandarrows.com/making-emotional-connections-through-participatory-design/ [Accessed 19 November 2017].

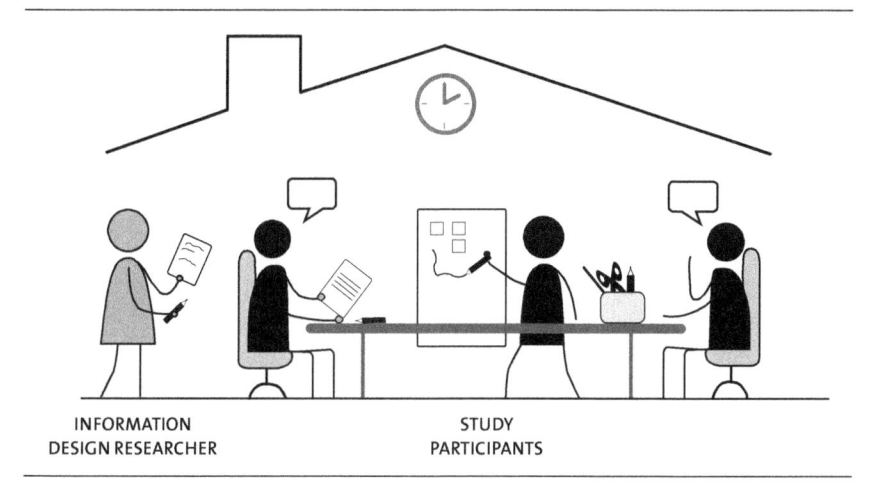

COLLABORATIVE WORKSHOP SETUP

Figure 5.10
When conducting a collaborative workshop, you will go to intended audiences' or stakeholders' workplaces. Participants will create some form of artefact to communicate their thoughts and dreams or engage in activities to generate and analyse ideas.

printouts with words or images directly related to the topic. For example, if exploring participants' feelings towards a specific product, you can bring photographs of the product and of other competitors' products.

- **Storyboard activities.** Creating storyboards can help participants articulate steps involved in their current activities, in a journey or a process; describe a series of events; or imagine ideal experiences from beginning to end. Participants draw or write a short description in sequential order of each step or moment that is relevant to them. Many options can be used for this technique. One way is to design storyboard templates to guide participants, without being prescriptive, and also give them generic collections of icons, images, and symbols. These help participants express their feelings and emotions, and include details in the storyboard. Another way is to give participants storyboards with some pre-defined elements or frames— for example, if working on better understanding a specific process and you would like participants to only focus on those steps that are less clear or need support.

- **Rapid prototyping activities.** Creating physical models helps participants communicate ideas or initial concepts. Prototypes can be simple sketches representing the structure of something (e.g. wireframes) or physical mock-ups created with paper, Lego, wood, or plastic. Encourage participants to sketch ideas on large sheets of paper so they can add to or comment on the initial idea. This technique is often used when working in the design of a tangible solution, such as the organization of information in an information graphic, or creation of an interface, as each participant can draw different structures, features and elements separately and then, as a team, assemble them to build an ideal layout.

2. Create outline workshop and instructions. After you have determined the generative activities, create a workshop outline and clear instructions to explain the flow of the session and help participants understand tasks and exercises. Pay attention to the choice of words, the phrasing, and the images used. It is important to pilot the flow of the workshop and the activities to check whether allocated time frames are appropriate to complete any given tasks and make sure instructions are clear.

3. Determine instruments and materials. To support activities, create instruments like worksheets, printouts with words or images, or templates. There is also a varied range of materials you can bring to a session to support each task, including everyday materials for creating prototypes (e.g. wooden bricks, Lego or building blocks, plasticine, pieces of rope), scissors, markers, glue sticks, stickers, sticky notes, big sheets of paper, and recording equipment (notetaker, or video/audio). The more materials you bring, the more you help participants feel unconstrained to externalize and share their ideas and thoughts.

4. Find location. While a collaborative workshop can be organized in any location, as far as possible, request a spacious, bright room with ample workspace, a big table, and empty walls where participants can pin things up.

Participants

The number of participants you recruit for these workshops can vary, based on the study's goal. Sometimes, you may organize many sessions with very few participants, while at others, one session with a larger group. Typically, a workshop won't involve more than ten participants. Recruiting participants with different backgrounds is important, but avoid recruiting friends. For example, if facilitating a workshop in a company to learn about the workplace dynamics, recruit people from different departments and backgrounds (e.g. developers, marketers, legal staff, consumers, etc.).

Data collection

A collaborative workshop involves the following parts:

1. Introduction. Explain the goal of the session and of the study, unless this could bias participants' performance, in which case you can explain this goal at the end of the session. Provide an overview of the activities for the day and clear expectations, and conduct one or two short warm-up activities to help participants get to know each other and feel more comfortable.

2. Generative activities. Introduce each activity indicating specific tasks participants should complete, time frames, and whether they should use any specific materials.

- **Document.** Capture and record everything generated by participants during each session, regardless of the chosen activity: take photos of sketches, drawings, intermediate steps, and final artefacts. If possible, also keep prototypes, sketches, and anything participants generate. Take notes of what participants are doing, of any moments when they seem to struggle or enjoy the activity.

3. Show & Tell. Ask participants to share with the rest of the group what they did and explain why they made those decisions. This will help participants articulate their ideas and thoughts.

4. Open discussion. End the session with a discussion about the activities, emerging ideas or concepts, and reflection about each participants' process. This will help you better understand participants' thinking process and latent needs.

Figure 5.11 provides a visual summary of how to plan and conduct collaborative workshops.

Online field research

The Internet has increasingly been playing a key role in making field research more accessible to information design. Some advantages, compared to face-to-face field research are: customizing methods, spreading their use while maintaining the necessary rigour to gather high-quality data[22], opti-

---

22 Hanna, P. (2012) Using internet technologies (such as Skype) as a research medium: A research note, *Qualitative Research*, 12(2), 239–242; Flick, U. (2009) *An Introduction to Qualitative Research*, 4th edn, Thousand Oaks, CA: SAGE.

HOW TO PLAN AND CONDUCT A COLLABORATIVE WORKSHOP

1. PLAN

DEFINE GOAL
What do you want to get out of the collaborative workshop?

DEFINE GENERATIVE ACTIVITIES
What will be the most appropriate activity?
- Collage Activities
- Storyboard Activities
- Rapid prototyping Activities
- Other

CREATE WORKSHOP OUTLINE & INSTRUCTIONS
How long will participants need to complete each activity?

DETERMINE INSTRUMENTS AND MATERIALS
What are the materials needed?

FIND LOCATION

2. PREPARE

Create workshop outline and instructions

Create activity briefs

Bring diary or notebook

Bring materials based on activities: pens, pencils, papers, scissors, sticky notes, etc.

Bring video recording device (e.g. smartphone)

Recruit and coordinate with participants

3. TEST BEFORE GOING INTO THE FIELD

PILOT STUDY
How can you ensure that you will collect the data you need?
- If you encounter problems, revise ① or ②
- Otherwise, start study in the field

4. GO INTO THE FIELD

Introduce study and conduct warm-up activities

Explain generative activities, and document and capture

Show and tell

Facilitate open discussion

Figure 5.11
Key steps to plan a collaborative workshop.

mizing data collection, saving time for transcription, and providing access to remote participants. Virtual data helps gain a more holistic understanding of people's needs, values, and preferences, and insights gathered from field online research can be used to triangulate data gathered from traditional field research methods.

You can transfer and adapt established qualitative methods to make them compatible with online needs, but access to strong and reliable Internet, and learning how to work with conference calls, design blogs, or sites are imperatives. Some examples are *virtual ethnography*, *online interviewing*, *online collaborative sessions*, and *digital self-documentation studies*[23]. The first two methods are discussed in the next sections.

Virtual ethnography

Virtual ethnography[24] has become common in design fields, particularly when working in multicultural and international projects. Using specific own tools, such as screen capture software, this mode of research adapts observation methods to study the digital space by exploring online culture and communities, and capturing interactions. These studies focus on understanding the modalities and contents of Internet communications and the textual ways in which a specific group and sub-group interact. The focus can be, for example, groups with specific interests in large social networks like Facebook, Reddit, or Twitter. Because you don't have to be physically in the same space as these groups to observe what they are doing, the "field" can be accessed remotely from any computer or place in the world.

In information design, you can use this method to learn and better understand people's mental models and experiences[25]. For example, to gain a better understanding of patients' experiences in hospitals of a specific city you can analyse their social media posts on social platforms, blogs or discussion boards, and any newspaper articles published on the topic. Another example would be if you have to design a visual explanation of a complex process, such as "how to get a green card in the US". The analysis of online forums will help you identify unclear steps in the process and discrepancies between what official sites communicate and what really occurs, and understand the diversity of issues applicants struggle with the most. These insights will help determine what type of information would be more helpful for the intended audience to be displayed in a visual explanation and with what level of detail.

23 ibid.
24 Flick, U. (2009) *An Introduction to Qualitative Research*, 4th edn, Thousand Oaks, CA: SAGE.
25 Mika, M. (2016) The big(ger) picture: Why and how virtual ethnography can enhance generative design research, *Medium* [online], Available at: https://medium.com/sonicrim-stories-from-the-edge/the-big-ger-picture-why-and-how-virtual-ethnography-can-enhance-generative-design-research-c39acb92fe4e [Accessed 20 January 2018].

Gathering data: methods for exploratory studies | 111

Online interviewing

Online interviewing is useful when you have an international or geographically hard-to-access sample, and it may be too expensive and time-consuming to meet with each participant in person. The question-answer structure in an online interview is the same as in a traditional interview, but two interviewing forms are available:

- **A synchronous online interview:** You and the participant connect online at the same time and use the same virtual space. When both are connected, you conduct the interview in the traditional way. A quiet place and a strong Internet connection on both sides are necessary; take into account any time differences.

- **An asynchronous online interview:** You and the participant aren't online at the same time; instead, you send your questions first and they return their answers after an agreed time. This exchange is often done via email. Avoid sending all questions at the same time; rather, send one or two questions first, and only send more questions after you have received the first answers. This avoids participants feeling interrogated or overwhelmed about responding to a long list of questions. Give participants a specific amount of time or deadlines to send their responses, to maintain control of the interview. Before commencing the email exchange, prepare, and send to participants consent forms and written instructions, which clearly explain the study and what they must do. Ensure you receive these signed consent forms from each participant before you start sending questions.

Participants' responses in face-to-face or synchronous interviews are more spontaneous, while responses in asynchronous interviews are more thought-through, as participants have more time to reflect on each question.

The methods discussed here can also be used to assess a design concept or evaluate a prototype in their context of use. The next chapter provides guidance to conduct these types of evaluation studies.

6 Gathering data: methods for evaluation studies

Perhaps, as many information designers, you conduct "informal" studies to validate concept ideas, or evaluate designs in progress among the members of your team, colleagues, or friends, by asking their thoughts on a design you are working on; this is a good starting point. However, these suggestions, even if useful, don't offer a rigorous or credible way to measure the effectiveness of a design or how an intended audience would use those designs in their context of use. In other words, insights from friends or colleagues can't be used as evidence in a meeting with a client or stakeholders. Furthermore, these insights don't represent the actual intended audience's feelings and points of view about the designs.

Instead, formal *evaluation studies* are needed to gain confidence in a design and elicit more accurate data that the team can use to move forward in the information design process, and that can be presented to clients as compelling evidence of its *performance* and that the design has potential to work as intended. The same methods—observations, contextual interviews, contextual inquiry, etc.—discussed in the previous chapter can be used to conduct these evaluation studies.

Information design evaluation dimensions

In information design, *performance* is related to whether the design supports the intended audience's cognitive activities, improving the quality of their work and experience to achieve their goals[1]. Typically, the role of evaluation is to measure how much the performance of a new design has increased, in comparison with that of an older version or, if no previous version exists, to measure whether the design achieves the performance expected[2,3]. Several aspects within the information design process can impact this overall

---

[1] Pontis, S. & Babwahsingh, M. (2016) Start with the basics: Understanding before doing, in *VisionPlus 2015 Conference: Information+Design=Performance Proceedings*, (IIID, IDA, Birmingham, England), 90–102; Frascara, J. (ed.) (2015) *Information Design as Principled Action: Making Information Accessible, Relevant, Understandable, and Usable*, Champaign, Il: Common Ground Publishing LLC.
[2] ibid.
[3] Lawson, B. (2005) *How Designers Think. The Design Process Demystified*, London: Architectural Press.

performance. Therefore, if time and budget permit, design work should be evaluated at different moments in the information design process:

- at the **conceptual design stage** to establish whether the concept idea is well received by the intended audience (Concept design step),

- when there is a **functional prototype**, to determine whether features and functions work as planned (Detail design step),

- and at a later stage, after the **prototype has been implemented**, to monitor whether it is working as intended (Evaluation step).

At each of these three points, four dimensions are often used to assess information design performance:

1. **What people feel and think about a design.** *Satisfaction* can be measured by eliciting people's opinions about an idea, the appropriateness of a design concept, or their experiences while using a functional prototype.

2. **How people use a design and the quality of what they produce.** *Usability* can be measured by analysing how people use a design in the intended context of use and examining whether its characteristics, functions, and features are used as intended and help complete the tasks they were designed to support. This dimension can be measured by assessing the learnability, usefulness, efficiency, and effectiveness of a design. Another way to measure usability is to examine the overall quality of the results produced before and after using a design (*outputs*).

3. **People's behavioural changes after using a design.** *Behavioural changes* can be measured by studying intended audiences' habits and analysing how successful they have been in achieving their goals since using the new design for some time (*outcomes*). This dimension is measured at the end of the design process by quantifying the improvement of something after the implementation of a design such as whether there is a reduction of effort in performing a task, time decrease in achieving a goal, or increase in the efficiency of a device[4].

4. **Whether people come away with a correct (or good enough) understanding after using a design.** While the previous dimensions help measure the effectiveness of a design in terms of outputs and outcomes, they can't determine whether users have gained the correct understanding. In some cases, users may show a change in behaviour, but it may not necessarily be the intended change that the design was

---

[4] Frascara, J. (ed.) (2015) *Information Design as Principled Action: Making Information Accessible, Relevant, Understandable, and Usable*, Champaign, Il: Common Ground Publishing LLC.

aiming for. Users' understanding from a design can be measured by assessing actions.

Each dimension—satisfaction, usability, behaviour change, and action—can be studied independently through qualitative or quantitative evaluation studies, but information design performance is more accurately measured by combining different types of methods. Examples of evaluative questions for each dimension are included in Table 6.1. As an example, observational studies can be used to qualitatively measure users' satisfaction and outputs' effectiveness (e.g. Are users engaging with the design? What do they feel about the design? Are the outputs they generated with the design in line with what was intended?). Usability can be assessed by quantitative methods that quantify the quality of what users have achieved after interacting with a design, or by qualitative methods: studying how users interact with a new design, their satisfaction with a design, and the experience and quality of what they did. Longitudinal studies are appropriate to assess changes in behaviours after a design has been implemented and used for a while, and determine whether the changes were the intended ones.

These are just a few examples of the various combinations of methods used to assess information design performance. To holistically assess the quality of information design solutions, triangulating qualitative insights and quantitative evidence is essential, as well as assessing a design at various stages of development and conducting evaluation studies in contextual settings.

Assessing a design at various stages of development

Depending on where in the information design process you are, the level of design fidelity of your prototype, and the goal of the evaluation, you can design two types of evaluation studies: *formative* or *summative*.

Formative evaluations

This type of evaluation focuses on exploring how an intended audience receives an idea or concept, or how a design can be improved while still being developed. No specific number of studies should be conducted, but the more feedback and input you can gather on an idea throughout the design process, the more robust the final version will be. Formative evaluations can be added to the process as short and frequent field studies that fit within a small budget and focus on examining whether the project is addressing the intended audience's needs. For example, each formative evaluation could

SATISFACTION **What people feel and think about a design**	USABILITY **How people use a design and the quality of what they produce**
EVALUATIVE QUESTION EXAMPLES	
• How does the intended audience perceive the idea? • How does the intended audience perceive the final design? • How did intended audiences describe their experience after using the solution?	• What functions of the new design did intended audiences use as planned? • Did intended audiences find the new design clearer? • Was the design easy to use? • Did the design help intended audiences achieve their goals?
COMMON ATTRIBUTES & MAIN GOALS	
Preference: • To determine whether users have a strong preference for old designs that may affect the use of a new design	**Efficiency:** • Determine whether information can be accessed in a shorter time • Determine whether the design helps fully accomplish intended goals in a shorter time
Opinions: • Identify users' opinions gained from looking at a design	**Quality:** • Identify user-perceived quality • Determine whether outputs satisfy initial needs and have increased quality • Determine whether the tasks are appropriate to complete using the design
Feelings: • Identify users' feelings and attitudes about the solution and its use	
	Accessibility: • Determine whether any specific features are needed to increase the design ease of use • Determine whether the design can be used by people with disabilities
	Learnability: • Determine whether the design is easy to remember how to use • Determine whether both frequent and infrequent users can successfully understand and use the design after some time of familiarizing with it
	Expertise: • Determine whether users need a specific baseline knowledge to understand and use the design
	Usefulness: • Determine whether the design enables users to achieve their goals and the specific goals for which the solution was created
POSSIBLE METHODS	
• Contextual interviews • Observations • Contextual inquiry • Collaborative workshops	• Contextual inquiry • Collaborative workshops • Diary studies • Design probes • *Usability testing*

Table 6.1 Information design performance dimensions specifying key attributes that can be tested individually and examples of evaluative questions for each dimension. Methods in italics are often used as part of quantitative studies.

OUTCOMES
People's behavioural changes after using a design

ACTIONS
Whether people come away with a correct understanding after using a design

- Did the new design reduce the problem?
- How successful has the design been in supporting users in achieving their goals?
- Does the design improve user performance as measured along dimensions like accuracy, speed, or quality?
- Are people making more informed decisions?

- Did the user accomplish the task as intended?
- Did the user change their behaviours as intended?
- Did the user show evidence of having gained the correct understanding after the use of the design?
- How many users have started to go through the process correctly?

Behaviour:
- Determine whether there has been a behavioural change after the user has used the design for some time

Accuracy:
- Determine whether the goal of the design was correctly understood
- Learn how long information is retained, regardless of speed, after using the design
- Determine whether users' behavioural changes are the intended ones after using the design

---

- Contextual interviews with participants and key informants
- Observations
- Diary studies
- Design probes
- *Before- and after-design studies*

- Contextual interviews
- Diary studies
- Design probes

Gathering data: methods for evaluation studies | 117

test specific performance attributes or design features. While some evaluation is better than none, as far as possible, designs should be evaluated at different levels of completion, rather than only at the end, when there is a functional prototype. Formative evaluations are often qualitative studies, involving intended users' feedback, such as contextual interviews or collaborative workshops, but others, such as *heuristic evaluations* or *cognitive walkthroughs*[5], don't involve users. These latter evaluations are highly common in interactive information design projects, such as the design of apps or interfaces.

Summative evaluations

This type of evaluation focuses on better understanding how a highly functioning, near-complete prototype or finished design is used. Their goal is to determine whether something actually works in the intended way, by measuring its performance or effectiveness: that is, if the design helps users achieve what they want. Insights from summative evaluations can help optimize, refine, and enhance a design solution in subsequent iterations. Depending on the goal, summative evaluations can be quantitative or qualitative studies, conducted in either the field or experimental settings. Learnings from these studies can inform future design revisions or be the start of a new project.

Assessing a design with field evaluations

Although testing designs in their context of use is increasing, more traditional evaluation studies in information design involve formative and summative evaluations conducted in the designer's studio. *Before- and after-design* and *usability testing* are two common types of these evaluation studies.

Before- and after-design (or pre-test/post-test design). This method involves measuring a design at various steps in the information design process and then comparing the results (e.g. comparison of outputs generated with and without a new design) *before* and *after* its improvement[6]. The impact of an intervention (in this case, the redesign) is often measured in experimental settings (so, in a lab or designer's studio) to facilitate managing controlled

---

5 Nielsen Norman Group provides excellent guidance on how to conduct heuristic evaluation and cognitive walkthroughs. E.g. Nielsen, J. (1995) *10 Usability Heuristics for User Interface Design* [online], Available at: https://www.nngroup.com/articles/ten-usability-heuristics/ [Accessed 10 January 2018].
6 Sless, D. (2008) Measuring information design, *Information Design Journal*, 16(3), 250–258.

and uncontrolled variables (e.g. time to complete tasks, selection of tasks)[7]. Results are more reliable when the interval between measuring old and new designs is short, because the longer you wait to measure again, the more participants can be influenced by other factors unrelated to the design (e.g. context, state of mind, level of familiarity) and therefore show misleading final results. In some cases, key informants' input (Chapter 5) is used to assess participants' outputs created before and after the intervention. For that, key informants should follow a pre-determined evaluation guide. Participants' outputs are sent to them, together with the guide, or assessed on the spot during an interview.

Usability testing. Rather than comparing old and new designs, in these evaluation studies, participants are recruited to complete a set of pre-determined tasks that simulate realistic scenarios, using a given design. While this method can be used with a qualitative approach, usability evaluations are often conducted in experimental settings to allow more control over variables and pre-determined metrics that can then be compared across participants and analysed quantitatively. An example of this type of study is when participants are recruited to come to a design studio and work in a specific room (the controlled environment) with a newly designed interface to perform specific tasks while they are being observed, either by someone present in the room or from outside the room through a mirrored window. Learning whether these tasks are successfully completed, or any unpredicted issues emerge (e.g. labels aren't understood by users because the terminology is incorrect), helps make final improvements in the design before moving into production.

As explained earlier, in experimental settings, participants interact with the design in an artificial context, which doesn't reflect how they would use it in real life. Having another person in the room or knowing that they are being observed can also influence their decisions or make them feel uncomfortable. Some participants may want to finish the tasks as soon as possible, not paying much attention to the actual interface. In other words, testing a design in a lab may provide misleading results, as intended audiences may have very different behaviours when using the same design in their natural environments.

Assessing information design performance dimensions involves more than analysing numbers about usage and satisfaction. While numbers indicate whether a design functions as planned or not, they don't shed light on the reasons why, even when a design works as planned, the intended

---

[7] Oppenheim, A.N. (1992) *Questionnaire Design, Interviewing and Attitude Measurement*, London: Continuum.

HOW TO CHOOSE A FIELD EVALUATION

INFORMATION DESIGN PROCESS STAGE — **YOUR ROLE**

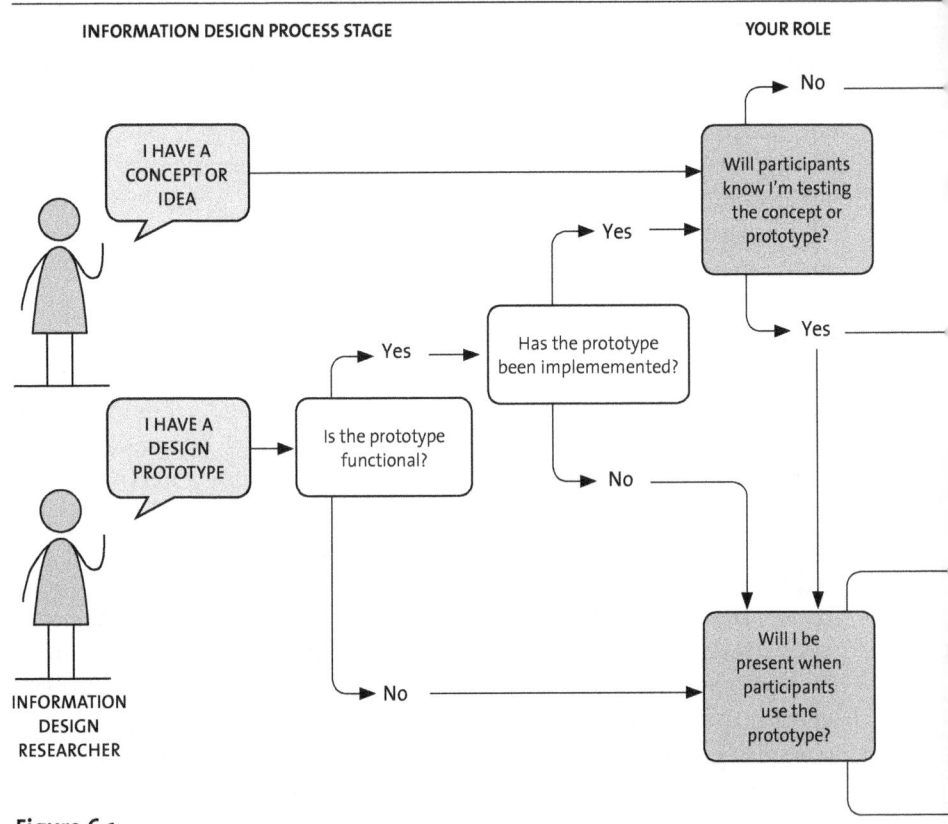

Figure 6.1
Decision tree summarizing the four types of evaluations for testing a design concept or prototype in the field.

audience doesn't seem to be engaged with it. A qualitative measure helps determine how usable something is or why a design didn't work[8]. Insights also help minimize assumptions and personal preferences (yours and your client's). Real-life nuances and issues are hard to replicate in experimental settings or may be unknown and only emerge when a design is used in its real context.

Field evaluation studies help gather these types of insights and reveal flaws in a potential direction or a functional solution because they place people at the centre by looking at how designs may be or are used in real-life

[8] Rubin, J. & Chisnell, D. (2008) *Handbook of Usability Testing: How to Plan, Design, and Conduct Effective Tests*, Chichester: John Wiley & Sons.

Figure 6.1

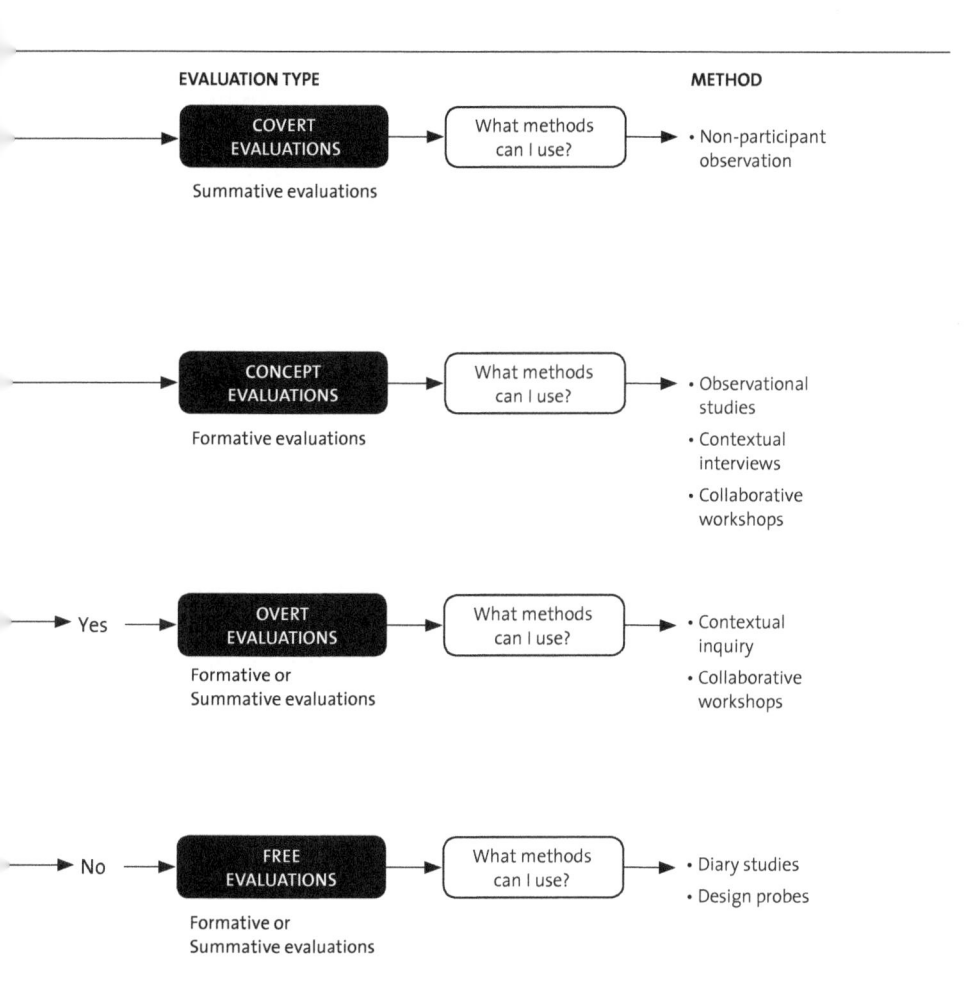

EVALUATION TYPE		METHOD
COVERT EVALUATIONS Summative evaluations	What methods can I use?	• Non-participant observation
CONCEPT EVALUATIONS Formative evaluations	What methods can I use?	• Observational studies • Contextual interviews • Collaborative workshops
Yes → **OVERT EVALUATIONS** Formative or Summative evaluations	What methods can I use?	• Contextual inquiry • Collaborative workshops
No → **FREE EVALUATIONS** Formative or Summative evaluations	What methods can I use?	• Diary studies • Design probes

situations. Using this type of evaluation, you can study many attributes that together provide a picture of how a design addresses each performance dimension, as indicated in Table 6.1. Not all attributes can be addressed with only one method or evaluation study, but certainly you must determine in advance those you plan to focus on. This helps determine the most appropriate evaluation approach and methods you can use to assess designs in their context of use. The following four approaches combine methods discussed in the previous chapter and can be used to conduct formative or summative evaluations: concept evaluations, covert evaluations, overt evaluations, or free evaluations. Figure 6.1 can help you decide what type of evaluation would be more appropriate depending on the information design stage and quality of the design solution you need to evaluate.

Concept evaluations

Purpose

The goal of concept evaluations is to assess interest and the appropriateness of an idea or the specific features of an idea with intended users early in the development process. The focus is on gathering participants' emotions, feelings, reactions, and opinions about the concept, as well as new ideas. Insights can be used to make radical changes to an initial concept or improvements on specific aspects of an idea. *Contextual interviews*, *observational studies*, or *collaborative workshops* can be conducted as part of this type of formative evaluation.

Preparation

As when preparing contextual interviews to gather exploratory data, first prepare an interview guide. Topics or questions may be related to aspects of the concept idea or specific features you are trying to learn more about.

Another important step is creating a simple model or prototype that visualizes the idea. Having a prototype during the study helps gather valuable insights, as participants have something concrete to react to and can better understand the idea. A prototype can be anything that takes a physical form: rough sketches, a wall of sticky notes, a role-playing activity, an object, an interface mock-up, a storyboard.

Participants

Recruit participants that correspond to intended users' characteristics. If you are testing a digital concept or product, it is important to determine the level of expertise (e.g. expert, intermediate, novice) and any specific skills (e.g. information technology skills, use of social media, use of database software, etc.) that participants need to have to fully understand the design and its fuctionality.

Data collection

Follow steps and guidance provided for *Observational studies*, *Contextual interviews* and *Collaborative workshops* in Chapter 5.

Covert evaluations

Purpose

The goal of covert evaluations is to observe intended users' interactions with a new design in a real setting with no guidance given, to assess how the new design is being used, and whether the content, features, and functionality are used and understood as intended. The focus is to better understand how well a new design works when people are distracted, interrupted, in noisy, crowded places, and in normal situations for them. Observed participants' behaviours can confirm initial learnings or indicate unexpected ways of using a new design. The main method is *non-participant observation*.

In information design, these evaluations help to assess new designs used in public spaces, like maps or wayfinding systems in hospitals, airports, or museums; or ATM interfaces in supermarkets or the street. As an example, in a map design project, after a newly redesigned map of the London underground was implemented and used for a few months, the design team observed commuters' interactions with the new maps at underground stations at different times of day. This type of formative evaluation was also an essential step in the Case Study 2: Legible London discussed in PART IV. Covert evaluations were conducted to test people's interactions with a new wayfinding system at different times of the day (morning rush hour, quiet afternoons), and on different days of the week to examine more accurate behaviours, intrinsic to a busy city like London. These studies helped assess feasibility and make sure that the new wayfinding system was working as intended. They pinpointed areas that worked well and others that needed further development.

Preparation

The first step is to define an observation guide. The bulk of this should focus on understanding the interactions between intended users and the new design but also include directions to identify any external factors that may impact those interactions. Another step is gaining familiarity with the location where the observations will occur: what will be a good place for you to observe? What are the specific characteristics of each time slot in the location? Are mornings more crowded than afternoons? For example, to observe whether a wayfinding system works in a shopping centre, you need to observe the setting when it is crowded and when only a few people are around. Create an observation schedule based on these identified time slots, days, and locations.

Participants

For covert observational studies, requesting informed consent may not be practical or may disrupt the normal flow of activities. However, to avoid ethical issues, limit data gathering to what you can observe during the study. When reporting insights, ensure you remove any data that may identify the people observed.

Data collection

Follow the steps and guidance provided for *Observational studies* in Chapter 5. If possible, conduct observations at different intervals over a few days to gather varied and rich data.

Overt evaluations

Purpose

Overt evaluations are a form of usability testing conducted in the field[9]. Their goal is to assess how well the new design works, by observing participants use it in their locations (home, workplace) and using their own equipment and tools, and to identify pain points for the participant or unexpected flaws in the new design. Often, participants won't be alone in the house or devoting their full attention to using the design. They may have children interrupting them, colleagues walking in and out of the office, ringing phones, or they may get distracted by some other task. Paying attention to how the new design fits in this context provides the depth of insight to develop a more successful output. After participants have completed a few tasks, interview them to gain a clearer understanding of the experience; they can also be questioned while completing a task. In contrast to conventional usability testing, you can suggest tasks for participants to complete using the new design, but they choose the order and ultimately decide whether those tasks are appropriate for them. Another way of using this approach, similar to contextual inquiry, is changing roles with the participant: you assume the role of apprentice and the participant explains how to use the new design. The main method is *contextual inquiry*, but you can also organize a *collaborative workshop*.

Portable designs, like small appliances, instructions or tools, or those participants can access from home or workplace (apps, websites, online tools),

---

9 Nielsen refers to usability testing in the field as flexible user tests. Available at: https://www.nngroup.com/articles/field-studies/ [Accessed 12 January 2018].

can be tested with this approach. Insights help reveal tacit aspects of participants' work practice, such as motivations, workarounds, and strategies that they may never articulate but which structure their work. You can conduct overt evaluations as a formative evaluation to find problems early on and improve the design or as a summative evaluation to validate the design against specific goals, identify strengths and weaknesses, and assess effectiveness.

Preparation

As the study starts with a contextual interview, firstly create an interview guide. This should be open and flexible so that participants can be involved in the construction of the interview. Questions should focus on uncovering participants' knowledge about, experience (if any) of, and intentions for the new design.

Secondly, create a *test guide* to give the study structure, and have a clear roadmap of the specific dimensions it should cover. One way to create this guide is listing core tasks the new design should support and specifying the initial order (e.g. by complexity, priority, length of completion) in which they should be completed. These tasks and the specified order are your starting point, but they may change from participant to participant. This is an important difference from usability testing: in overt evaluations you observe a participant move between tasks and decide what to do. As sessions progress, you improvise and customize sets of tasks to better suit each participant's needs. If you have assumed the role of apprentice during the study, the test guide should also include a list of the specific things that a novice user would need to learn about the design to know how to use it.

A functional prototype of the new design is essential. Depending on whether it is a formative or summative evaluation, the prototype will be more or less complete or functional. Prepare cards, with an explanation of each task, to give participants. These can be designed cards or a sheet of paper; give one task at a time.

If video recording the session, ensure you have the necessary equipment, for example video camera, eye-tracking cameras, or screen capture software.

Participants

Although the sample is small, participants should be representative. Based on the design being tested, determine whether participants should work in a certain industry or require technological or specific knowledge or experience to successfully interact with the new design.

Data collection

A basic structure for overt evaluations is contextual interviews followed by participant observation sessions and closing contextual interviews or conversations. Techniques for gathering data are the same as those for exploratory data gathering explained in Chapter 5.

During the initial interview, point out to participants that you aren't testing them but the design. It should be clear that you are evaluating the functionality of the design and not their work. During the observational session, don't become defensive if participants don't understand the design and things aren't working as intended; instead, at the end, ask participants to unpack their feelings and explain any unclear aspects of the design.

Free evaluations

Purpose

The goal of free evaluations is to ask participants to use and evaluate the new design in their own time. Participants use the new design as part of their lives for an agreed period to complete given tasks and then report the experience during an interview. As part of the study, participants are given a list of tasks to complete, but they choose how and when to do them.

This type of evaluation is less structured than overt evaluations but demands a higher level of commitment from participants. The methods often used are *diary studies* or *design probes*. Portable designs, like mobile phones, apps or tools, or those participants can access from their home or workplace, can be tested with this approach, which is also useful when intended users live remotely. For example, for a free evaluation study to test the usability and effectiveness of a tool in its context of use, at Sense Information Design we prepared a self-documentation package to help participants have a better sense of what type of data they should document during the study, but they were free to use the tool to support any relevant task. Recruited participants didn't live in the same country, so we mailed the package in advance to them and then we discussed it during a Skype interview to kick off the study. Throughout ten working days, participants uploaded visual and written data to a shared Dropbox folder. Some unplanned issues occurred, like one of the participants unable to go to work and therefore capture data for two days. However, by the end of the assigned period, both participants gathered rich high-quality data. The study helped identify aspects for improvement in the tool and determine whether and how it was supporting the intended users' job.

Preparation

As with exploratory self-documentation studies, most of the planning for this evaluation concerns the logistics and design of the package and tasks. Tasks aim at studying participants' use of the new design in their natural habitat and own time. They should be[10]:

- **Challenging.** Participants must achieve a goal (e.g. create something, book a flight, buy a new product).
- **Realistic.** Tasks represent a real-life situation.
- **Appropriate for the participants.** Tasks are the kind that would be often encountered in participants' environments.
- **Not too large.** The complexity of the tasks is in line with the time available.
- **Feasible in the time available.** Tasks involve using the new design to achieve a goal that can be observable, to serve as the basis for worthwhile analysis.
- **Suited to background.** The complexity of the tasks is in line with participants' experience.

As with the previous two types of evaluations, a functional prototype that participants can use and interact with is necessary. If a prototype involving technology isn't fully operable, unimplemented technology can be simulated, for example with the *Wizard of Oz* technique[11]. This technique involves a human or team, unknown to participants, simulating some or all of the responses generated by the new design, in order to make it look functional to participants. The *Wizard of Oz* technique can be used to test mobile apps, web interfaces, and other computer-related applications. For example, working with a team, we used this technique as part of a diary study to help test an app prototype that had reduced functionality[12]. Human wizards simulated the behaviour of the app, but to minimize the risk of affecting users' freedom of expression, we told them that they would be interacting with a computer system and we didn't mention the wizards. During ten days, 20 participants interacted with the app and recorded their experiences us-

---

10 Cross, N., Christiaans, H., & Dorst, K. (1996) *Analysing Design Activity*, Chichester: John Wiley & Sons.
11 Dahlbäck, N., Jönsson, A., & Ahrenberg, L. (1993) Wizard of Oz studies—why and how, *Knowledge-Based Systems*, 6(4), 258–266.
12 Pontis, S., Kefalidou, G., Blandford, A., Forth, J., Makri, S., Sharples, S., Wiggins, G., & Woods, M. (2016) Academics' responses to encountered information: Context matters, *Journal of the Association for Information Science and Technology*, 67(8), 1883–1903.

ing the prototype. After this period, we interviewed participants to better understand the recorded data and their experiences with the app, and identify areas for improvement.

Participants

The sample can be small but with representative participants willing to commit for the length of the study.

Data collection

Free evaluations follow the same data collection structure as self-documentation studies described in Chapter 5. As with overt evaluations, point out to participants that you aren't testing them but the design. Of equal importance is stressing the role of documenting everything they do and experience while using the new design (e.g. things they like, things they don't understand, things they would change), and generating evidence of what they do (e.g. take photos, record themselves, draw sketches, video tapes).

The next step in the process is to start making sense of collected data, and Chapter 7 provides direction and support to complete this task.

7 Making sense of field data

The previous two chapters examined various methods of gathering rich and diverse qualitative data. Now the focus is on making sense of that data.

Regardless of whether you gathered qualitative data to better understand people's needs or to test a new design, making sense of field data involves many tasks, the following being the core ones: giving some sort of order to the data, distinguishing relevant from irrelevant data, identifying significant patterns, and finding a way to communicate what the data revealed and how that informs design decisions and actions. While there is no recipe or unique way of performing these tasks, an important consideration applies to all qualitative analysis: **get familiar with data first; look for categories later; generate meaning last.** Be rigorous and systematic; don't skip steps, even if you think you know the story after reading half the data or you are running out of time. Initial thoughts and ideas are likely to be assumptions. And even if you are correct, you aren't allowing the possibility of finding unexpected insights that could later distinguish your design solution from others.

In qualitative analysis, when you start making sense of data, you look for insights to:

- Learn new details and information.

- Test initial ideas.

- Break assumptions and misconceptions.

- Confirm what you and your team already knew.

- Highlight important things that you and your team didn't know about the design but should know, to move forward or improve the design.

This chapter provides guidance and direction, as well as methods to support the analysis process but is by no means exhaustive or confining. Adapt what is presented here to fit your specific situation and study.

Understanding sensemaking

Expanding the definition of senemaking introduced in Chapter 1, this process involves two complementary activities: **analysis** and **interpretation**, followed by **communication**.

Analysis

Analysis is the examination of individual pieces of data to deepen understanding. Through analysis, you bring order to data and organize it into patterns, categories, and basic descriptive units. This task in qualitative research fundamentally involves becoming fully immersed in the data, by combing through it with an *interpretative lens* to identify themes and patterns. While each qualitative approach to data analysis has its particularities (e.g. grounded theory, thematic analysis, cross-case analysis), they all involve some type of **coding**. Coding involves assigning labels, representing symbolic meaning, to words, 'chunks', or segments of the data gathered during the study[1]. These labelled words or segments are your *codes*. A code groups *ideas or concepts*, related to the original goal of the study, that share similar meaning.

Coding can be done solo or as a team. Each member of the team can code and analyse the data separately and then compare and discuss, or you can code and analyse the data together[2]. Later in this chapter, the "Generate codes" section provides a detailed explanation of how to code data and examples to illustrate each step.

Interpretation

Once data has been coded and themes and patterns identified, the next stage involves describing, giving meaning to, and interpreting those codes, categories, patterns, and themes in context; that is, in relation to the study's initial goal and what you want to learn. *Interpretation* involves the reconstruction of the findings from the analysis into new ways to answer "why" questions, and reveal stories, patterns, themes, relationships, and connections, previously unknown. Simply put, interpretation attaches meaning and significance to the analysis to explain findings and identify linkages among descriptive dimensions.

Your goal is to attach meaning to what you found, provide an explanation, make inferences, draw conclusions, and apply lessons to other contexts. Generating meaning from someone's actions, feelings, and behaviours involves dealing with 'unobservables' rather than 'observables'[3]. Your inferences and conclusions connect the two. Interpreting data can feel uncomfortable and overwhelming at first, particularly if you have never done it before or are used to market research or other quantitative data analysis methods.

---

1 Miles, M.B., Huberman, A.M., & Saldaña, J. (2013) *Qualitative Data Analysis*, SAGE.
2 Patton, M.Q. (2002) *Qualitative Research & Evaluation Methods*, 3rd edn, Thousand Oaks, CA: SAGE.
3 Miles, M.B., Huberman, A.M., & Saldaña, J. (2013) *Qualitative Data Analysis*, SAGE, 292.

While many methods can be used to aid analysis, interpretation happens in the mind: you interpret and recombine ideas, describe patterns and themes that emerged from the analysis, and explore ways to distil them in a way that reflects new understanding of the subject or problem. This is the challenge of interpretation: how to minimize the subjective quality intrinsic to this activity. As generating meaning can vary from one person to another, minimizing bias and ensuring credibility is essential (Credibility and Transparency criteria – Chapter 3). Ways to achieve this are describing any cases that show the opposite of what your explanation argues, highlighting any irregularities in the data and mentioning other views. Furthermore, it is preferable to interpret data with your team rather than alone. Various interpretations stimulate dialogue and generate richer insights. Many interpreters may notice things that you alone may not. If interpreters have different backgrounds (e.g. information designers, researchers, programmers), they may approach data from different points of view; this is expected in this type of qualitative analysis, and interpreters should discuss differences and disagreements until they find complementary, but not contradictory, interpretations and meanings in the data[4].

In practice, sometimes these two activities, analysis and interpretation, can occur in parallel or in repeating cycles, and there is no clear distinction between one ending and the other beginning. In some types of qualitative research, like grounded theory[5], these cycles can occur simultaneously with data collection, as the study moves forward. Here, each cycle is discussed separately.

Communication

While interpretation occurs inside your head, *synthesis* is a way to share findings with others. The main goal is figuring out the underlying structure or overarching explanation that best represents or describes findings in relation to the subject, phenomenon, situation, or problem at hand. This requires finding a way to communicate or represent connections among patterns and themes. That is, you need to construct a framework (e.g. a story, a theory), for communicating the essence of what the data reveals that illustrates connections between themes and provides explanations and recommendations in relation to the study's goal.

---

4 Blandford, A., Furniss, D., & Makri, S. (2016) Qualitative HCI research: Going behind the scenes, *Synthesis Lectures on Human-Centered Informatics*, 9(1), 1–115.
5 Patton, M.Q. (2002) *Qualitative Research & Evaluation Methods*, 3rd edn, Thousand Oaks, CA: SAGE.

Sensemaking step-by-step

Following data gathering, you have different types of sets of field data. This data is often unstructured, as it isn't organized in terms of a specific set of analytical, pre-determined categories. As shown in Figure 7.1, sensemaking begins with the preparation of your data to ensure that whatever you find at the end of the process is credible and valid and, most importantly, informs the information design process of the project at hand.

Prepare data

These four steps can help you prepare your data for analysis:

1. Sort. What type of datasets have you gathered? How much did you gather? Become familiar with what you have. Organize datasets by some logic, for example by participant, by method, or by data type, so you can create an *inventory*. For each dataset:

- Include date of data collection, location, and who was involved.

- Anonymize participant's name, if you haven't already done so. Remove participants' personal information from all datasets. Anonymize data before printing out any notes and information that could identify participants, such as names, towns, project titles, company names, etc. To be consistent, use the same conventions employed while collecting data.

- Indicate characteristics (e.g. number of pages) and type (textual or visual, handwritten or digital), and number pages.

- If you are working with digital files, indicate location by route and folder name. Location is particularly important and helps datasets to be reachable when you start the analysis. If you are sorting a pile of handwritten field notes, clearly label each pile.

The inventory helps determine which gathered datasets are high quality and worth analysing, which would be best to start the analysis with, whether they connect to each other, or whether to follow a specific order.

2. Check. Is my data useful? As shown in Table 4.2, not all qualitative data collected is complete or high quality. It is important to:

- **Check for completeness:** This is related to whether you have gathered similar amounts and types of data for each participant. For example, if you only have visual data for one participant out of ten, you may decide to exclude that dataset from the analysis or to try to gather visual data from the other participants.

SENSEMAKING PROCESS FOR FIELD DATA

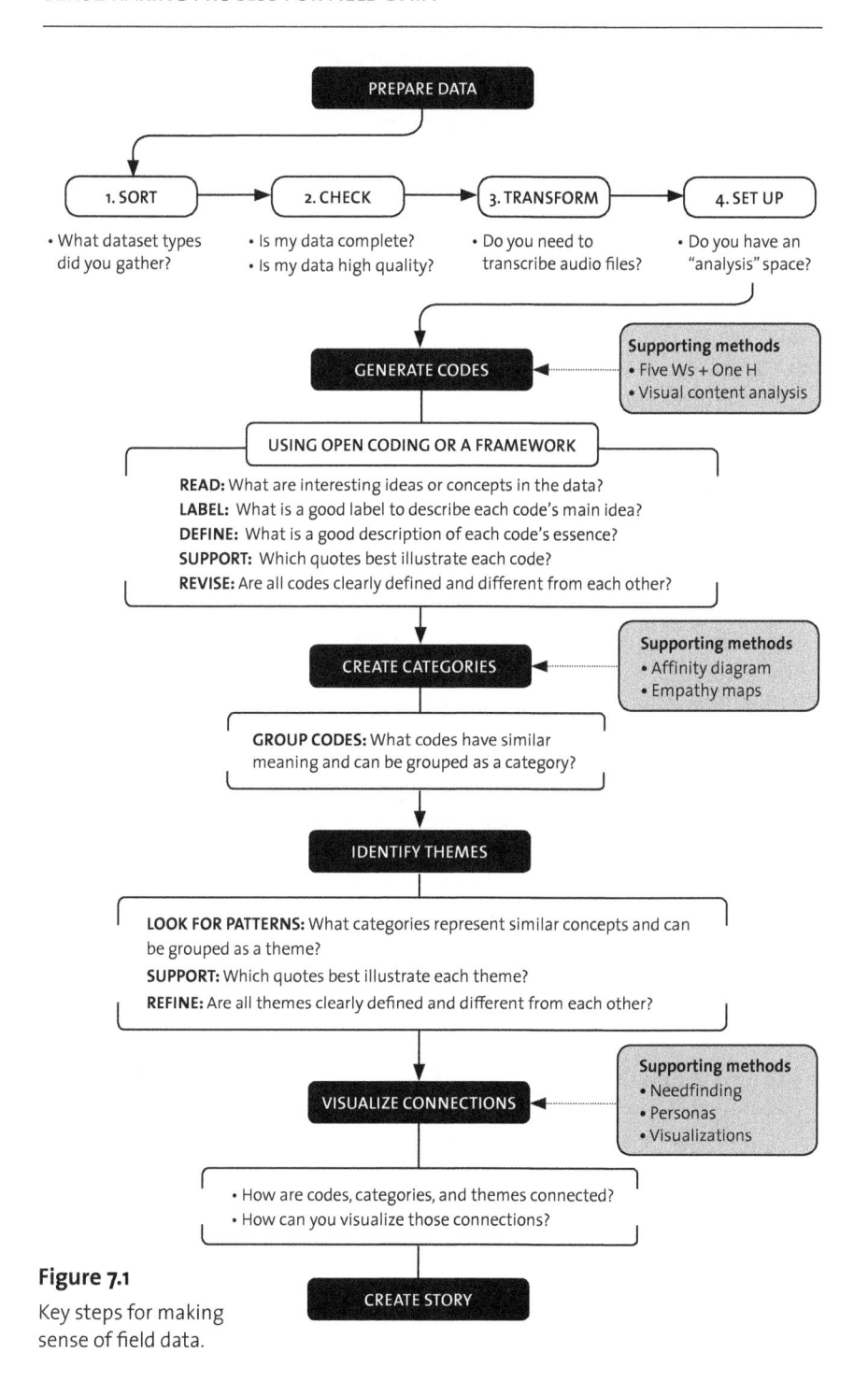

PREPARE DATA

1. SORT
- What dataset types did you gather?

2. CHECK
- Is my data complete?
- Is my data high quality?

3. TRANSFORM
- Do you need to transcribe audio files?

4. SET UP
- Do you have an "analysis" space?

GENERATE CODES

Supporting methods
- Five Ws + One H
- Visual content analysis

USING OPEN CODING OR A FRAMEWORK

READ: What are interesting ideas or concepts in the data?
LABEL: What is a good label to describe each code's main idea?
DEFINE: What is a good description of each code's essence?
SUPPORT: Which quotes best illustrate each code?
REVISE: Are all codes clearly defined and different from each other?

CREATE CATEGORIES

Supporting methods
- Affinity diagram
- Empathy maps

GROUP CODES: What codes have similar meaning and can be grouped as a category?

IDENTIFY THEMES

LOOK FOR PATTERNS: What categories represent similar concepts and can be grouped as a theme?
SUPPORT: Which quotes best illustrate each theme?
REFINE: Are all themes clearly defined and different from each other?

VISUALIZE CONNECTIONS

Supporting methods
- Needfinding
- Personas
- Visualizations

- How are codes, categories, and themes connected?
- How can you visualize those connections?

CREATE STORY

Figure 7.1
Key steps for making sense of field data.

- **Check for quality:** This is related to whether you have elicited relevant and rich data from each participant. If participants' responses are monosyllabic, lack a descriptive layer, or are wordy but completely unrelated to the purpose of the study, the resulting data is of poor quality. Poor-quality data is hard to analyse because there is little to compare with other participants, find patterns, or draw conclusions. High-quality data has a mixture of nouns, adjectives, and verbs, representing emotions and experiences, even if sentences are grammatically incorrect.

If, after completing these first two steps, you feel you don't have enough rich data, don't try to make it work. In some cases, it may be necessary to spend one or two more days in the field and collect more data, and use the first study as a pilot. Ultimately, this is more productive than trying to make poor-quality data work.

3. Transform. How accessible and usable is the data? Determine what data needs to be transformed into a format that is easily accessible by you and all team members, and easier to work with. This can involve:

- **Transcribing audio or video files:** Unless you have the budget and planned to send audio files to be transcribed by a professional, performing a full transcription can be time-consuming, as it typically takes four to six hours to transcribe an hour of audio, depending on the level of detail required. If you don't have the luxury of time (and budget), just transcribe selected parts, directly relevant to the study topic.

- **Digitalizing field notes:** Type field notes as a digital document, free of abbreviations or acronyms. As you re-read your notes, and memories from the study come to mind, you can expand these with details that you previously didn't note down. Clean and legible notes invite comments and facilitate coding and analysis, even when working solo.

- **Annotating videos:** You can use specialized software to add labels to specific parts of your videos, or you can print out specific frames or parts and add any annotations by hand.

- **Printing digital photos or other digital files:** If analysing manually, print out various copies of the anonymized data; keep one as the master copy and return to it to check data in context and locate quotes and details. Use another copy for annotations, comments, and coding data. If you have a very large dataset, you could print out selected extracts that you consider surprising or revealing.

4. Set up. Where can I analyse the data? It is hard to analyse data virtually because you don't have full view of the material. Rather qualitative data is analysed outside the computer. As mentioned in Chapter 4 when you started the design of the study, with full view of the material on one or several continuous surfaces, you can easily scan across notes and pages of information and zoom in and out as needed to see details or broader patterns. This enables visual association, as opposed to stacking and piling documents, which enforces linear, page-by-page analysis[6]. In your designated study room:

- **Maximize the "analysis" space:** Spread out all your notes and datasets on a table and/or pin them up on the walls in an organized manner so you can track the origin of each data extract.
- **Or post materials neatly on panels:** If you don't have a designated room for the study, pin up your notes and datasets on foam board panels or flipchart sheets of paper.

Now you are ready to start making sense of the data. The next sections provide a step-by-step guide to this process, following *thematic analysis*[7]. This is a commonly used method for analysing qualitative data, which provides a useful framework for information designers to make sense of field research data.

Generate codes

Analysis starts with coding data to generate codes. This involves thorough reading, asking questions of the data, and looking for answers, through labelling words, phrases, excerpts, and images that have similar meaning. At the beginning, it may be difficult to identify key ideas or concepts from people's descriptions and interactions, and you may have to read the same data more than once. Figure 7.2 describes two ways of coding data: **using open coding**[8] or **using a framework or roadmap**[9], and provides examples of questions you can use to code data using the former technique.

---

6 Miles, M.B., Huberman, A.M., & Saldaña, J. (2013) *Qualitative Data Analysis*, SAGE; Chipchase, J. (2017) *The Field Study Handbook*, 2nd edn, Field Institute.
7 Clarke, V. & Braun, V. (2014) Thematic analysis, in *Encyclopedia of Critical Psychology*, New York: Springer, pp. 1947–1952.
8 Given, L.M. (ed.) (2008) *The SAGE Encyclopedia of Qualitative Research Methods*, Thousand Oaks, CA: SAGE.
9 While direct interpretation is often followed, *interpretation models* are commonly used in design contexts as frameworks to structure analysis. This type of model is a set of agreed topics or categories for interpretation, defined in advance, based on the study's goal. *Interpretation topics* are used to structure collected data and start the coding process. Mattelmäki, T. (2008) *Design Probes*, 2nd edn, Vaajakoski, University of Art and Design Helsinki.

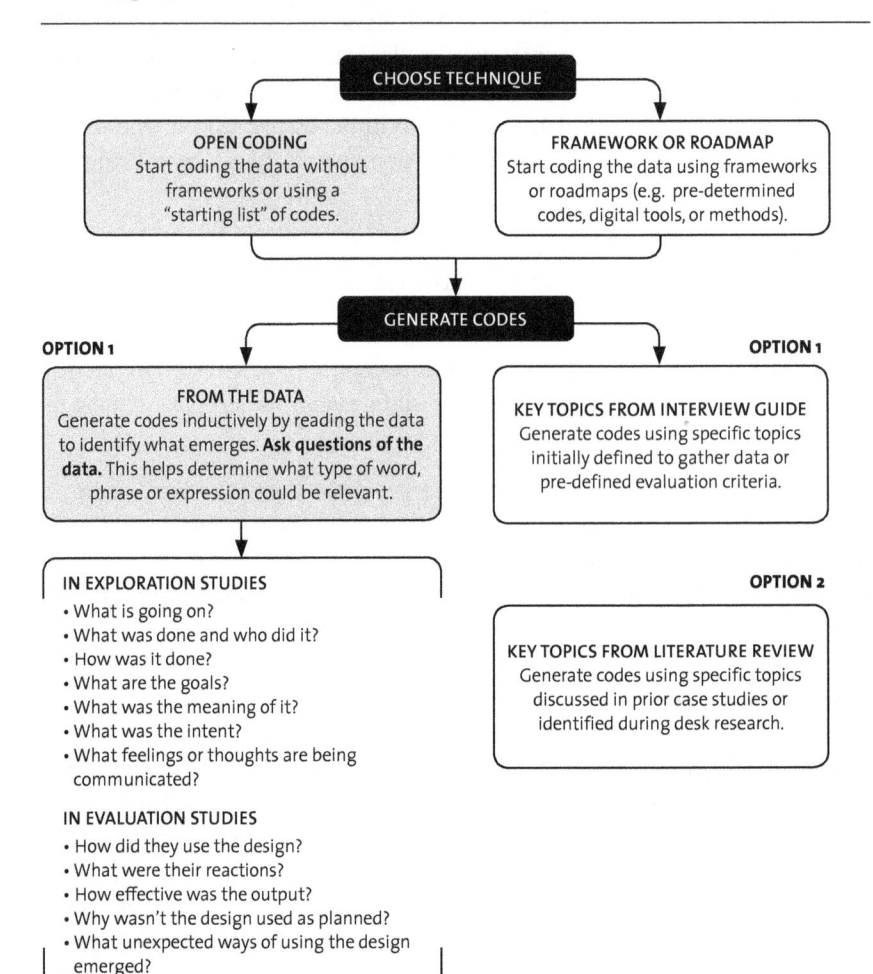

TECHNIQUES TO CODE FIELD DATA

CHOOSE TECHNIQUE

OPEN CODING
Start coding the data without frameworks or using a "starting list" of codes.

FRAMEWORK OR ROADMAP
Start coding the data using frameworks or roadmaps (e.g. pre-determined codes, digital tools, or methods).

GENERATE CODES

OPTION 1

FROM THE DATA
Generate codes inductively by reading the data to identify what emerges. **Ask questions of the data.** This helps determine what type of word, phrase or expression could be relevant.

IN EXPLORATION STUDIES
- What is going on?
- What was done and who did it?
- How was it done?
- What are the goals?
- What was the meaning of it?
- What was the intent?
- What feelings or thoughts are being communicated?

IN EVALUATION STUDIES
- How did they use the design?
- What were their reactions?
- How effective was the output?
- Why wasn't the design used as planned?
- What unexpected ways of using the design emerged?

OPTION 2

BASED ON PROFESSIONAL EXPERIENCE
Generate codes based on your initial understanding of the topic under investigation and using any questions that emerged during the field part of the study as a "starting list" of codes.

IN EXPLORATION & EVALUATION STUDIES
- What was the goal of the study?
- What do you think you will find?
- What do you remember from the interviews?

OPTION 1

KEY TOPICS FROM INTERVIEW GUIDE
Generate codes using specific topics initially defined to gather data or pre-defined evaluation criteria.

OPTION 2

KEY TOPICS FROM LITERATURE REVIEW
Generate codes using specific topics discussed in prior case studies or identified during desk research.

Figure 7.2
Open coding can be hard if this is the first time you code field data, but you can use questions for guidance with what to look for in the data.

Read data. First read the data line-by-line, looking for interesting ideas and concepts. These can be events, features, phrases, behaviours, values, feelings, or stages of a process, and can manifest as single words, phrases, excerpts, expressions, images, objects featured in photographs, or actions noted in videos. Also look for *unexpected* but interesting and recurrent ideas or concepts completely unrelated to the study. Look at the following excerpt:

> **RAW FIELD DATA EXCERPT:**
>
> "My family was not that physically active at all, I was encouraged to do sports sporadically. I was more involved in musical theater as a kid. Growing up my mom did not cook super healthy, stuff like mac n cheese, spaghetti, we always had vegetables though. The past couples of years she has had more healthy cooking habits. My dad also smoked growing up. My parents didn't really model healthy habits. I didn't start eating healthy and exercising until I left and went to boarding school in 9th grade".

Label codes. Clearly distinguish—pin up, number, colour, circle, stack, or annotate—each group of ideas or concepts that shares similar meaning to generate codes. For now, don't worry about whether or how the codes relate to each other. Attach a *descriptive label* to each code, summarizing its basic topic or idea. Code labels can be a descriptive word or short phrase, such as nouns (e.g. "Emotions", "Sports"), or representative phrases from participants' own words (e.g. "Eating healthy") or yours (e.g. "Early commuters"). Figure 7.3 presents three options you can use to label codes and examples of codes.

Define codes. Each code should be clearly and concisely defined, so you and other coders can consistently code the data, as indicated in Figure 7.4. When codes are well defined, 85% to 90% agreement is expected between multiple coders, depending on the number of codes. Having operational definitions of each code from the start helps minimize disagreements during the process and build reliability.

Support codes. Use participants' quotes or data extracts to exemplify each code. This helps determine whether all initial codes are valid. Highlight hard copies or literally use index cards or sticky notes to extract the sentences that illustrate a code's meaning. If you can't find quotes to support a code, discard the code. Sometimes, what originally seemed to be a code turns out not to be on a second reading. This is part of the process.

Revise codes. Initial definitions can change and be revised as the coding process progresses and you gain more familiarity with the data. As you go through the data a second time, you will notice that some codes repeat,

WAYS TO LABEL AND INDICATE CODES IN THE DATA

OPTION A: Numbering
Add numbers to indicate each code. Create a key with each code label.

(1) "My family was not that physically active at all, I was encouraged to do sports sporadically. I was more involved in musical theater as a kid. **(2)** Growing up my mom did not cook super healthy, stuff like mac n cheese, spaghetti, we always had vegetables though. **(3)** The past couples of years she has had more healthy cooking habits. **(2)** My parents didn't really model healthy habits. I didn't start eating healthy and exercising until I left and went to boarding school in 9th grade".

CODES:

(1) *Sports*
(2) *Role models*
(3) *Routine change*

OPTION B: Marking
Highlight, underline, or box words or sentences that denote a code. Add code label next to each code. You can use this option to code data manually or using Microsoft Word.

CODE: *Sports* → "My family was not that physically active at all, I was encouraged to do sports sporadically. I was more involved in musical theater as a kid. Growing up my mom did not cook super healthy, stuff like mac n cheese, spaghetti, we always had vegetables though". ← CODE: *Role models*

OPTION C: Extracting
Extract words or sentences that denote a code and write each on a sticky note or index card. Add the respective code label to each sticky note or index card. You can use this option to code data and create an affinity diagram.

"My family was not that physically active at all, I was encouraged to do sports sporadically. I was more involved in musical theater as a kid".	"Growing up my mom did not cook super healthy, stuff like mac n cheese, spaghetti, we always had vegetables though".	"The past couples of years she has had more healthy cooking habits".
CODE: *Sports*	**CODE:** *Role models*	**CODE:** *Routine change*

EXAMPLES OF CODE DEFINITIONS

CODE: *Sports*
CODE DEFINITION: Comments related to outdoor or indoor physical activities performed by participants

CODE: *Role models*
CODE DEFINITION: Comments related to other people's health awareness and their impact on participants

CODE: *Routine change*
CODE DEFINITION: Comments that indicate any sort of change towards a more healthy lifestyle

Figures 7.3 and 7.4

Examples of coded data, codes, and code definitions. Code labels can be one or two words. You can use your own words or the participant's exact words to label a code.

INITIAL CODE	OPERATION	REVISED CODE
Sports	Initial code was **renamed** to make it more specific	*Exercise*
Role models	Initial code was **divided** to make it more specific	*Family role model* *Friends role models*
Routine change	Initial code was **merged** with another code as it wasn't different enough	*Routine change* is now coded as *"Eating healthy"*

Table 7.1 Examples of operations you can use to revise codes.

while others are only mentioned a few times; some codes don't work anymore, other codes that you didn't see in the first reading emerge, and others are divided into two distinct codes (Table 7.1). As the analysis progresses, codes can be:

- **Renamed.** Sometimes, initial code labels aren't representative enough. Discuss with your team more accurate ways of describing code labels or different ways of explaining the scope of the code, based on new findings. First, identify and clarify the meaning of those codes that seem to be central to your analysis. This can also result in the definition of new codes and sub-codes.

- **Divided.** Sometimes, a code becomes too complex; too many segments are allocated to the same code, or it involves too-contrasting ideas. In these cases, you can decide that that code would be clearer if split into two distinct codes.

- **Merged.** The opposite can also occur, if you realize that two codes aren't as distinct as you previously thought and therefore would be clearer if combined.

Coding generates many individual codes associated with their respective segments of data. There is no specific number of codes you should aim for. For example, with large datasets (e.g. data from 20 participants), you can have around 40 codes, but 10 to 20 codes often emerge from smaller sets (e.g. data from ten participants). For example, after conducting 11 contextual inquiry sessions, each of around 60 minutes long, I initially generated 19 codes. Throughout the coding and analysis process, you and your team need to question each code's relevance and validity.

There isn't a hard rule for when to stop generating codes; typically, you continue until you can no longer identify any different or new ideas. In other words, stop when you reach *saturation*; this is the point where you see ideas and concepts start to repeat. Reaching saturation doesn't mean that every sentence of a dataset has to be coded.

Create categories

Once data is coded, the analysis focuses on working with coded segments and extracts, rather than original raw data, although you can return to it any time to check contextual elements or clarify meaning. Here, you combine individual codes into a smaller number of clusters or *categories*[10] that share similar meaning or characteristics. To form categories, look for connections and similarities running through and across individual coded segments, and group those with similar meaning as a category, as illustrated in Figure 7.5. Creating mind maps or diagrams, to map out connections between codes and categories, or creating affinity diagrams are good techniques to support this step and the next one.

Identify themes

In this step, reduce the number of different pieces of data by condensing them into smaller but more encompassing or explanatory ones: *themes*. Themes are a higher-level categorization, representing conceptual 'threads' or patterns that tie together codes and categories[11]. The creation of *themes* starts by combining codes and categories with overlapping meanings, shared by most or many of the study's participants. This involves various iterations and revisions, ensuring that codes and categories allocated to each theme are representative and clearly exemplify the essence of the respective theme. The process of combining codes into categories, and then into themes is illustrated in Figure 7.6.

Look for patterns. Return to the coded data and read all participants' excerpts and quotes extracted for each code and category, and determine whether there is a coherent pattern. Pay attention to correlations, connections, or trends between them, and whether these show any specific directions, fluctuations, or changes. The following are six different ways of looking for patterns[12,13]:

- **Trends:** How often does "the behaviour" occur? Does the "behaviour" increase or decrease? Is there fluctuation? How much or how little does it change?

- **Magnitudes:** What are the levels of "happiness"? How "happy" are they?

---

10 Given, L.M. (ed.) (2008) *The SAGE Encyclopedia of Qualitative Research Methods*, Thousand Oaks, CA: SAGE; Miles, M.B., Huberman, A.M., & Saldaña, J. (2013) *Qualitative Data Analysis*, SAGE.
11 Patton, M.Q. (2002) *Qualitative Research & Evaluation Methods*, 3rd edn, Thousand Oaks, CA: SAGE.
12 Babbie, E. (2010) *The Basics of Social Research*, 5th edn, Wadsworth Publishing.
13 Saldaña, J. (2015) *The Coding Manual for Qualitative Researchers*, Thousand Oaks, CA: SAGE; Miles, M.B., Huberman, A.M., & Saldaña, J. (2013) *Qualitative Data Analysis*, SAGE.

Figure 7.5
Categories created after combining codes from interview data. Princeton University students extracted interviewees' quotes and wrote each of them on a sticky note. Then they organized codes by similar meaning under five categories: stress, mindset, comparison, coping, and success.

- **Structures:** What are the different types of "tasks"? Does any of the tasks seem harder than others? Can "feelings" be ranked from highest to lowest?
- **Processes:** Is there any order among the "tasks"? Are they performed in any particular order?
- **Causes:** What are the causes of "happiness"? Does it occur often during "work interactions" or "social interactions"? Are there any strong or weak connections between "feelings" and "tasks"?
- **Consequences:** How does "work" affect "happiness", in both the short and long term? What changes does it cause in "behaviours"? Are there exceptions or outliers?

If a pattern is strong and representative enough, it becomes the essence of a theme. You can compare themes by looking for similarities and differences

FROM CODES TO THEMES

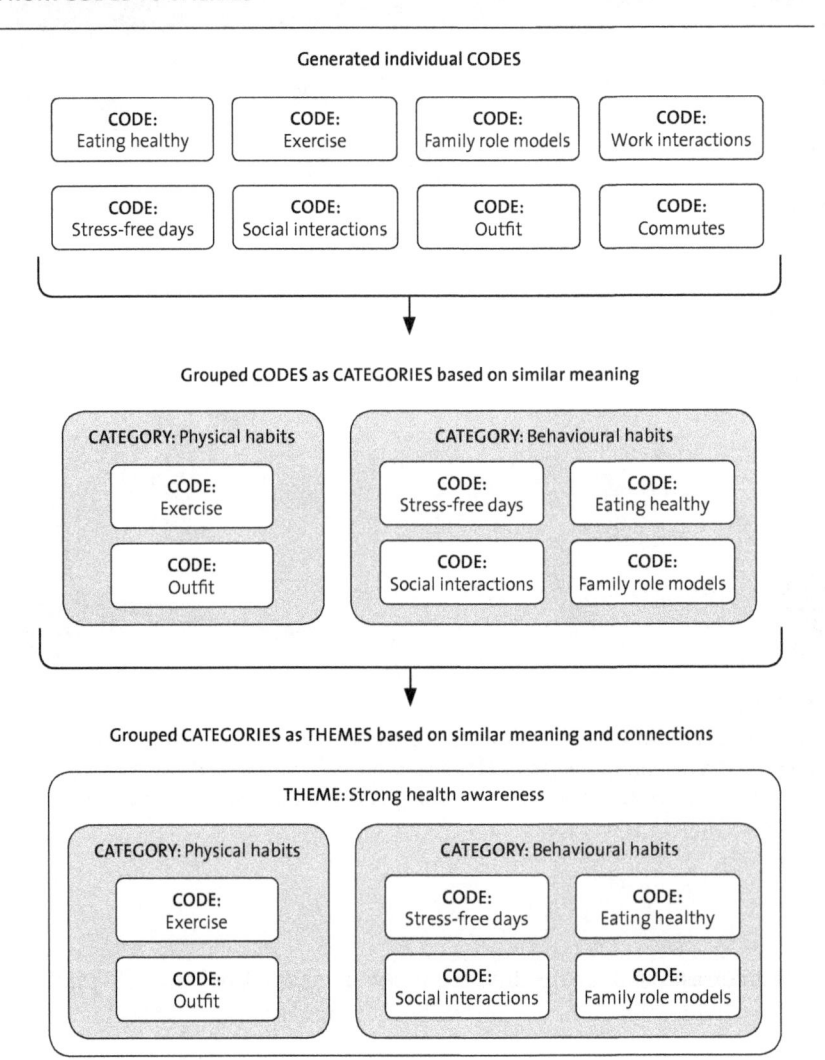

Generated individual CODES

- CODE: Eating healthy
- CODE: Exercise
- CODE: Family role models
- CODE: Work interactions
- CODE: Stress-free days
- CODE: Social interactions
- CODE: Outfit
- CODE: Commutes

Grouped CODES as CATEGORIES based on similar meaning

CATEGORY: Physical habits
- CODE: Exercise
- CODE: Outfit

CATEGORY: Behavioural habits
- CODE: Stress-free days
- CODE: Eating healthy
- CODE: Social interactions
- CODE: Family role models

Grouped CATEGORIES as THEMES based on similar meaning and connections

THEME: Strong health awareness

CATEGORY: Physical habits
- CODE: Exercise
- CODE: Outfit

CATEGORY: Behavioural habits
- CODE: Stress-free days
- CODE: Eating healthy
- CODE: Social interactions
- CODE: Family role models

Figure 7.6
Example of the coding process showing how codes are combined into categories and then themes using the same data excerpt and codes discussed in Figures 7.3 and 7.4.

or any connections between two or more of them, such as hierarchical relationships or correlations. With successive refinement and consolidation of codes, categories, and patterns, a clearer picture emerges of the predominant themes.

Support themes. As for codes and categories, use participants' quotes to exemplify and support each identified theme and pattern. To find compelling quotes, return to the raw data; don't only look at the coded excerpts or quotes. At the end of this step, sometimes, themes don't make sense or quotes don't adequately support a theme. In this case, you must decide whether to keep, discard, or modify the themes and relocate or discard the quotes.

Refine themes. Review and refine all themes to check whether they are relevant and sufficiently different from each other or whether some are too similar and therefore need to be combined or reordered. Give each a descriptive name that accurately represents its essence. Theme names need to be clear, concise, and immediately communicate what each is about[14]. If you feel themes are unclear or not representative enough, read and re-read the data to code any additional parts that you didn't initially notice. This helps enrich and clarify each theme. This step can be as long or short as you wish, depending on how many iterations with the data you perform.

You should end this step with a collection of themes and sub-themes, supported with extracts from the data. The number of themes varies from project to project. As a rough guide, if you have more than 14 themes, this often indicates that some can be merged into one or become sub-themes. Typically, in smaller studies, you end with three or four themes, while, in larger studies, it would be rare to have more than six or eight themes (the number of sub-themes could be much higher) at this stage. After each revision, the number of themes will reduce.

Visualize connections

Now that you have outlined a set of themes, start interpreting what you have coded and grouped. You must generate meaning or provide an explanation for why those themes emerged and are related, or others didn't emerge. Methods that support this step are needfinding, personas, and visualizations, which are explained later in this chapter

Then, look at one theme at a time to see whether there are links between codes, categories, and other themes to build a narrative or structure within each theme. Go back and forth between the themes and the data (evidence), and your own thoughts and views (your assumptions). Sometimes, you can start by testing your own assumption about the problem. This involves examining the data and considering whether themes support that assumption or provide alternatives. In other cases, it may be more helpful to connect participants' stories and themes to psychological or sociological

---

14 Clarke, V. & Braun, V. (2014) Thematic analysis, in *Encyclopedia of Critical Psychology*, New York: Springer, 1947–1952.

theories or cultural dimensions to see whether that reveals deeper meanings or explanations.

Pay attention to contradictions between what participants did and said. This helps identify tensions, issues, and unaddressed needs. Links between themes and sub-themes are often based on:

- **Hierarchies:** Connections based on ranks or levels, such as connecting "border control", "passport", and "immigrants" sub-categories to a "travels" category.

- **Processes:** Connections based on the sequence in which actions are performed, such as connecting "receiving form", "completing form", "submitting form" to "annual tax return".

- **Chronologies:** Connections based on temporal dimensions, such as "early morning activities", "late morning activities", "noon activities", and "afternoon activities".

- **Frequencies:** Connections based on how often something occurs, such as "never understand what to do", "sometimes understand what to do", "occasionally understand what to do".

Create rough visualizations to visually represent connections and meaningful structures between codes, categories, sub-themes, and themes, and how they relate to each other. The way you link themes and other components depends on your interpretation of the coded data and the study's goal. To better understand what each theme means, turn each name into a complete statement, summarizing its essence. The theme statements are your key findings, and the visualizations provide the context to describe the explanation behind each of them. Rephrase the findings as often as necessary until they communicate the idea clearly. Share findings with your team members to check clarity and meaning.

Create a story

This step's aim is to connect the final theme statements or findings to the study's goal, demonstrate how they address initial questions, and present any new insights. Any visualizations created in the previous step delineate the story's structure. The story should present a coherent understanding of what the data showed, describing how themes—the pillars of the story—connect to the initial goal, and highlighting findings in that same context. Chapter 8 provides guidance for creating the story, and discusses different formats for sharing findings.

The next sections introduce tools to support the sensemaking process and methods to support analysis and interpretation.

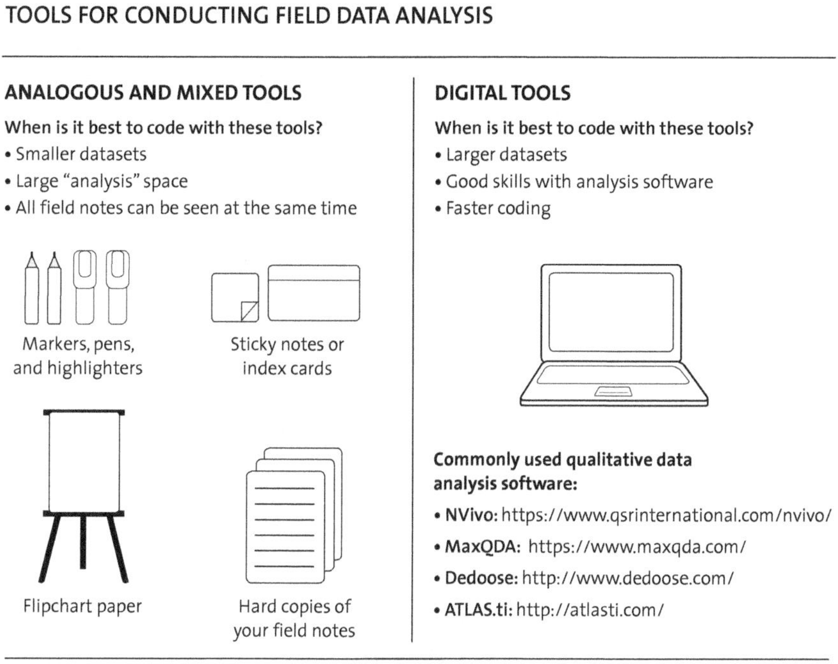

TOOLS FOR CONDUCTING FIELD DATA ANALYSIS

ANALOGOUS AND MIXED TOOLS

When is it best to code with these tools?
- Smaller datasets
- Large "analysis" space
- All field notes can be seen at the same time

Markers, pens, and highlighters

Sticky notes or index cards

Flipchart paper

Hard copies of your field notes

DIGITAL TOOLS

When is it best to code with these tools?
- Larger datasets
- Good skills with analysis software
- Faster coding

Commonly used qualitative data analysis software:
- **NVivo:** https://www.qsrinternational.com/nvivo/
- **MaxQDA:** https://www.maxqda.com/
- **Dedoose:** http://www.dedoose.com/
- **ATLAS.ti:** http://atlasti.com/

Figure 7.7

Tools for conducting field data analysis. Depending on the type and amount of field data you gathered from the field, you can analyse it manually or digitally.

Tools supporting analysis

Data can be coded manually, using analogous tools, or digitally, using specialized tools and software (Figure 7.7). Your experience, time, budget, access to appropriate technology, and size and type of datasets are factors to consider before deciding how to code the data. The greater the volume of data, the more helpful digital tools are; but some degree of familiarity with these tools is needed; otherwise, they may hinder rather than speed up the process.

Analogous and mixed tools

Any visual thinking tools are extremely useful to facilitate steps of the analysis process when working away from the computer. Office and design studio materials are the best tools to aid this process. Use them to tag, flag, and colour-code data to make differences, levels of emphasis, patterns, and connections easier to see. For example, use sticky notes or index cards to write important excerpts, quotes, or category labels. Or just highlight quotes, and write notes or category labels in the margins of transcript pages. Make the most of colour-coding: assigning a colour to each code or category, then using it consistently to code the data, helps easily identify different codes. Other common materials and their uses are:

- Markers and pens, for making notes and underlining
- Sticky notes, for adding comments or flagging important items
- Highlighters, for calling attention to different codes and pieces of data
- Sticky dots, for tagging related information or categories

Analysing qualitative data can be absorbing, and after a while it can be hard to remember every interesting learning or quote you see, or where they came from. To help track insights and learnings, keep a diary or index cards of your progress and thoughts.

During analysis, it's normal for work to get messy, but this messiness will help you "get inside" the content and start to see it from multiple dimensions. It is important to remember that analysis isn't simply a mechanical process of filtering bits and pieces—it's about uncovering important details or pieces of a larger story that come together through interpretation.

Digital tools

When datasets grow larger, manually organizing and coding them becomes significantly time-consuming. Then, using a *Qualitative Data Analysis (QDA)* tool to support the analysis process may be a better option. QDA tools can be used to code any type of data: visual, audio, and textual; they support data storage, retrieval, comparing, and linking. These tools mainly work by creating and applying a list of codes in a systematic way. This ensures coding consistency—essential when working with multiple codes—locating coded themes, grouping data together in categories, and comparing passages in transcripts or field notes to identify patterns. The larger the dataset, the more codes you have, some of which may overlap, as parts of a sentence could represent more than one code. In such cases, QDA tools can facilitate coding. For example, you could attach multiple codes to particular passages of data, with sentences or words overlapping one another and codes nesting within one another. Once data is coded, you could use these tools to export groups of coded quotations. While this could also be done manually, these tools minimize human error, such as mixing or forgetting codes. Similarly, if you are working with visual datasets, you could code visual material using digital asset management software like Adobe Bridge.

As new software is constantly changing, and new features are developed, the specifics of every available QDA tool are not discussed. Regardless of which QDA tool or software you decide to work with, don't try to learn and use every feature. If you are new to qualitative analysis, use digital tools to organize data, support coding, and thoroughly examine the coded data, to identify relevant themes and generate useful insights. Most importantly, while you could use many QDA tools to aid coding and analysis, the interpretation part of the process is on you. There is no QDA tool (so far) that will generate meaning from the data. Making sense of the data is your responsibility: decide what codes go together to form a pattern, what constitutes a theme, what to name it, what meanings to extract from datasets and case studies, how to tell the story. Software can help speed up the process, but it's no substitute for the power of your mind, your creativity, or intelligence.

Methods for organizing and coding data

The first steps in making sense of data involve organizing the material in some fashion and coding. While most qualitative studies follow some form of coding data, in professional practices open coding is less frequent. Instead, information designers often organize and code data, using methods, techniques, or frameworks that offer pre-defined dimensions or lenses. This section presents four of them as shown in Figure 7.8. But no single one does it all; each is meant to provide structure to help extract meaning from data and identify themes. After you have completed these activities, interpretation begins.

Five Ws + One H

Purpose

The goal of this framework is to organize and code data, using a list of 'basic questions' or *Five Ws + One H*[15]: who, what, when, where, why, and how or how much. Each can be used as a pre-defined code or category when reading through the data.

---

15 This method is often used in journalism for writing a news story or in police investigations as an interrogative method. Dan Roam (2013) introduced this method in the context of visual thinking as a way to identify the key components of a story that could be visualized. Flick (2009) discusses the use of these questions in qualitative analysis. Roam, D. (2013) *The Back of the Napkin: Solving Problems and Selling Ideas with Pictures*, London: Portfolio; Flick, U. (2009) *An Introduction to Qualitative Research*, 4th edn, Thousand Oaks, CA: SAGE.

METHODS FOR ORGANIZING AND CODING DATA

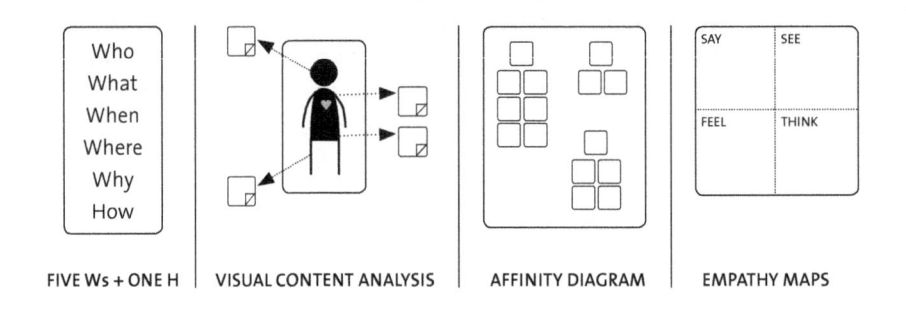

FIVE Ws + ONE H | VISUAL CONTENT ANALYSIS | AFFINITY DIAGRAM | EMPATHY MAPS

Figure 7.8
Four methods to support data organization and data coding. You can use them if you are analysing data either manually or digitally.

How to use it

Read the data and proceed through the basic questions one-by-one, picking out all occurrences of an answer as they appear in the data. For example, if analysing data about shopping patterns, you can formulate questions starting with those words to generate codes from different perspectives:

- **Who:** People involved and their roles, e.g. Who are the key figures in supermarkets (e.g. consumers, staff)?
- **What:** Issue or phenomenon, e.g. What are the major milestones for consumers (e.g. select products, go to checkout, pay)?
- **When:** Time, e.g. When do those major milestones happen (e.g. once a week, specific dates)?
- **Where:** Location, e.g. Where does shopping happen (e.g. farmers' market, local supermarket, deli, online)?
- **Why:** Reasons or intentions, e.g. Why do people go to get groceries (e.g. weekly shopping, forgot something, planning a party)?
- **How:** Strategies, tactics, steps, e.g. How did technological advances change the experience (e.g. online shopping, use of smartphones)?

Once all data is coded, sub-categories may emerge within initial categories.

Visual content analysis

Purpose

The goal of *visual content analysis*[16] is to code visual data, such as photographs, icons, maps, and drawings generated by participants at the end of a field session. Visual data is deconstructed into basic elements and features, and then labelled (coded) so that it has some analytical significance. Coded elements and features, within each visual dataset, are quantified to identify patterns and correlations between them. Then, coded elements and features (in each image) are interpreted by understanding how they connect to the wider context, within which the image makes sense. Figure 7.9 shows two examples of visual content analysis we did at Sense Information Design.

How to use it

1. Collect datasets. Collect a representative sample of images. Sample size depends on the amount of variation among all the relevant images. For example, for projects we did at Sense Information Design, we worked with relatively small but representative sample sizes: 22 photos for one project and 11 workshop outputs for another project. If working manually, in preparation for coding, print all images and pin them up on a wall, foam board, or spread them on a table. If working with material generated in workshops, scan it and print copies to work with during analysis.

2. Define categories for coding. Define categories for analysis based on the goal of the project. Categories should be objective and only describe what is on the images; that is, what you see. They can be interpretative or descriptive but have to be:

- **Exhaustive:** every aspect of the image with which the research is concerned must be covered by one category

- **Exclusive:** categories must not overlap

- **Enlightening:** categories must look at images through an analytical but interesting lens that is coherent with the study's goal

3. Code images using categories. When coding written data, you focus on finding interesting ideas and concepts manifested through words and phrases;

---

[16] Visual content analysis has its roots in content analysis methods, used to analyse written datasets and texts. Rose, G. (2012) *Visual Methodologies: An Introduction to Researching with Visual Material*, Thousand Oaks, CA: SAGE.

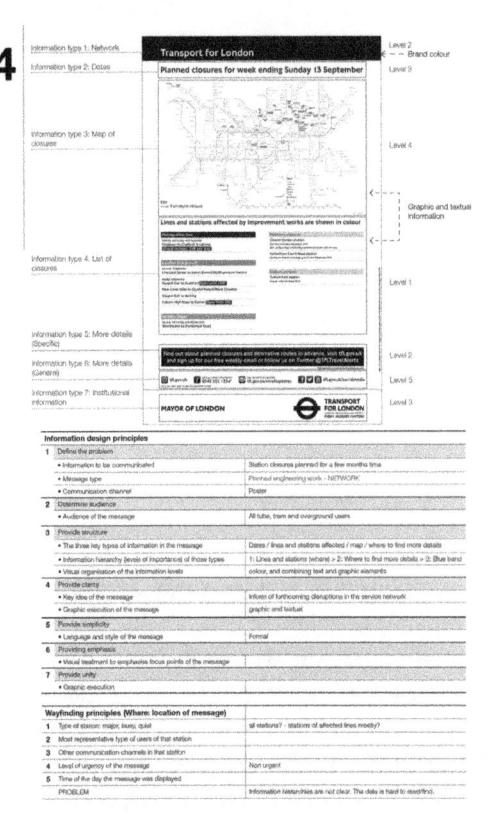

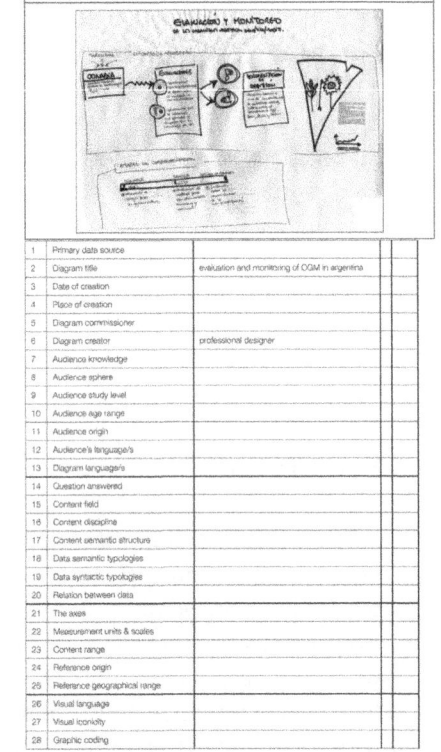

Figure 7.9

Visual content analyses. In the left image, the analysis focused on understanding whether information design principles were effectively used. Each element of the poster was categorized into seven information types. Posters were also analysed following wayfinding principles to determine whether their location and type of message were aligned. In the right image, the analysis focused on understanding outputs created by participants in workshop sessions. Similarly, component parts in each output were categorized following information design principles.

now, your codes are the details, elements, and features in each image. At the end of coding, each image should have several codes attached to it.

4. Analysis of coded images. One way of analysing coded images is to count the codes, to produce a quantitative account of the content, as I did in Figure 7.10 to better understand the graphic evolution of tube maps. This is complemented with analysis of the relationship between codes and other elements around them (context). In short, look at both the frequency of codes in each image and across images, and explore any connections between different coding categories.

Figure 7.10
Visual comparative analysis. The analysis focused on examining the evolution of the London Underground map from its creation in 1933 to its 2008 version. Component parts of each of these maps were coded, categorized, compared, and analysed.

Affinity diagrams

Purpose

The goal of *affinity diagrams*[17] is to help organize data. If you don't have a large dataset, you can use this method without coding the data first. Instead, you focus on identifying and extracting ideas and data, with similar meaning or referring to similar topics, and writing these on individual cards. Then individual cards with similar meanings (which represent codes or ideas) are sorted and clustered into groups, creating categories. Once all codes or ideas are grouped as categories, they are combined as themes based on related meaning. The last step is to examine the themes and categories for emerging patterns.

> 17 The KJ method was originally created by Jiro Kawakita in the 1960s to help categorize ideas, issues or solutions into a limited number of related groups. Kawakita, J. (1986) *KJ hou*, Tokyo: Chuokoronsha; Valsiner, J. & Rosa, A. (2007) *The Cambridge Handbook of Sociocultural Psychology*, Cambridge: Cambridge University Press.

Making sense of field data | 151

How to use it

This is a method, used without computers, which requires space to spread out the data, such as a large table, whiteboard, or wall.

1. Code data. Use the *Five Ws + One H* or open coding, following the steps described in the "Generate codes" section to code data.

2. Organize codes. Read through the coded data and:

- Extract, and write on individual sticky notes or cards, sentences, words, or expressions that exemplify each code (write big and clear, not small!).
- Put sticky notes or cards on a wall or whiteboard so you and each team member can see them all at the same time, as shown in Figure 7.11.
- Rearrange sticky notes or cards, putting those with related meaning and ideas together in columns or clusters.
- Repeat these steps until all codes are included in a cluster.

When working with a team, a possible starting point is for each person to choose the three ideas they find most interesting and put them up as the first elements of a cluster. Then each person looks for other sticky notes or cards with similar or related ideas. Codes representing ideas or observations that fit in multiple places should be written again for each cluster.

3. Identify themes. Once all sticky notes or cards are organized, read each column or cluster; combine any columns with similar ideas or meaning. Each resulting column or cluster is a theme. At the end of the process, you should have four to six main themes. Define a headline or concept that represents the main idea of each theme. If you have many more, re-read codes and categories within each theme to determine whether they are unique and distinctive enough. Too many themes often indicate overlapping meanings that can be merged.

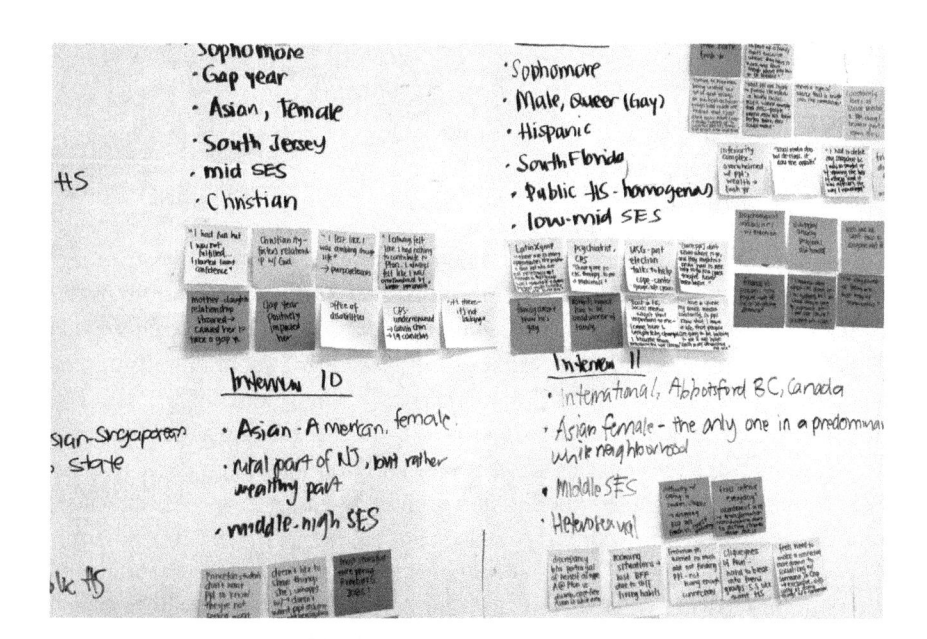

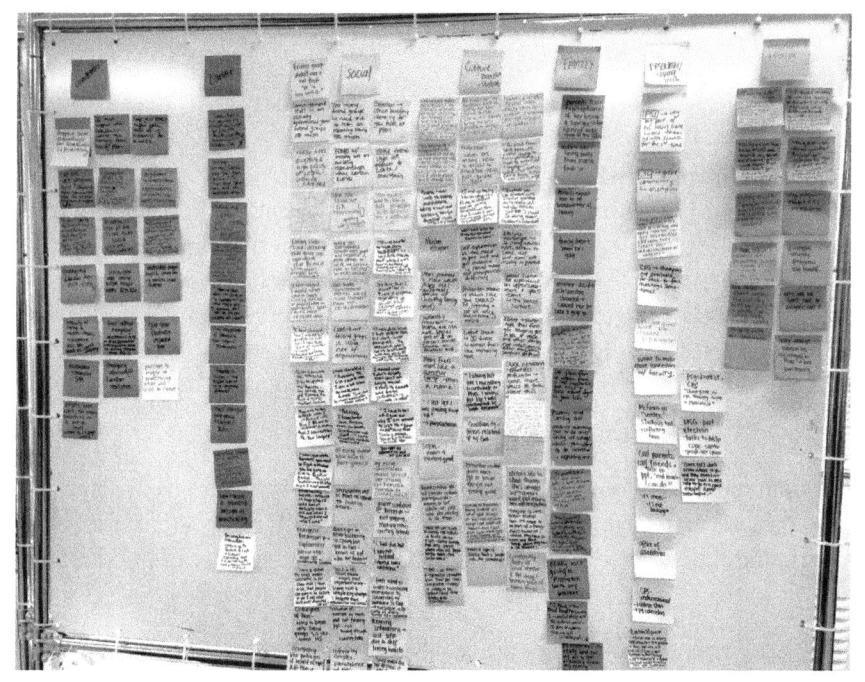

Figure 7.11
Affinity diagram to make sense of interview data. Princeton University students first organized codes per study participant (top image), and then re-organized the same codes into seven categories (bottom image). The name of each category is written on a post-it note at the top of each column.

Making sense of field data | 153

Empathy maps

Purpose

The goal of *empathy maps*[18] is to help identify different segments within a specific population or audience, and how each of them sees and experiences a given topic, situation, or problem. Each empathy map creates a picture reflecting the thoughts, feelings, and emotions about the problem, from each of those different segments. When used for data analysis, this tool helps code and categorize verbal and non-verbal cues by clustering them into (four, six, or eight) pre-determined activities: what people "say", "see", "hear", "think", "feel", what they "do", their "challenges" and "gains" regarding a topic. Empathy maps help cluster data for each activity and from across all participants of a study that show similar needs, goals, challenges, and ways of thinking about a given problem, even if the data doesn't come from the same person. You will probably create empathy maps as the first step, prior to creating *personas* (described in the next section). Empathy maps are typically created when working with other team members to gain a shared understanding of the intended audience.

How to use it

1. Create templates. First, create structures to populate each activity. Divide a large sheet of paper or whiteboard into as many sections as activities you want to explore; number and label each section with the corresponding activity. Numbering each section gives an explicit order to the sequence of activities your team will examine[19]. Think big: the size of each template should be big enough for each team member to write something or put a sticky note inside each section. Create at least three templates, to encourage an active session.

2. Populate activities. Populate each section with data from across participants that share similar ways of responding to each activity, regardless of their demographic characteristics, as shown in Figure 7.12. Use a set of pre-defined

---

18 Originally, this tool was created by Gray et al. (2010) to help imagine and identify the ideal user or customer for a product. You and your team would brainstorm, while walking through one entire day in the life of an ideal customer, imagining what this person would say, do, feel and what their characteristics would be. Consequently, this information would not be rooted in initial user research but on what you and your team imagined about an ideal user. Gray, D., Brown, S., & Macanufo, J. (2010) *Gamestorming: A Playbook for Innovators, Rulebreakers, and Changemakers*, O'Reilly Media Inc.

19 Gray, D. (2017) *Updated Empathy Map Canvas* [online], Available at: https://medium.com/the-xplane-collection/updated-empathy-map-canvas-46df22df3c8a [Accessed 19 November 2017].

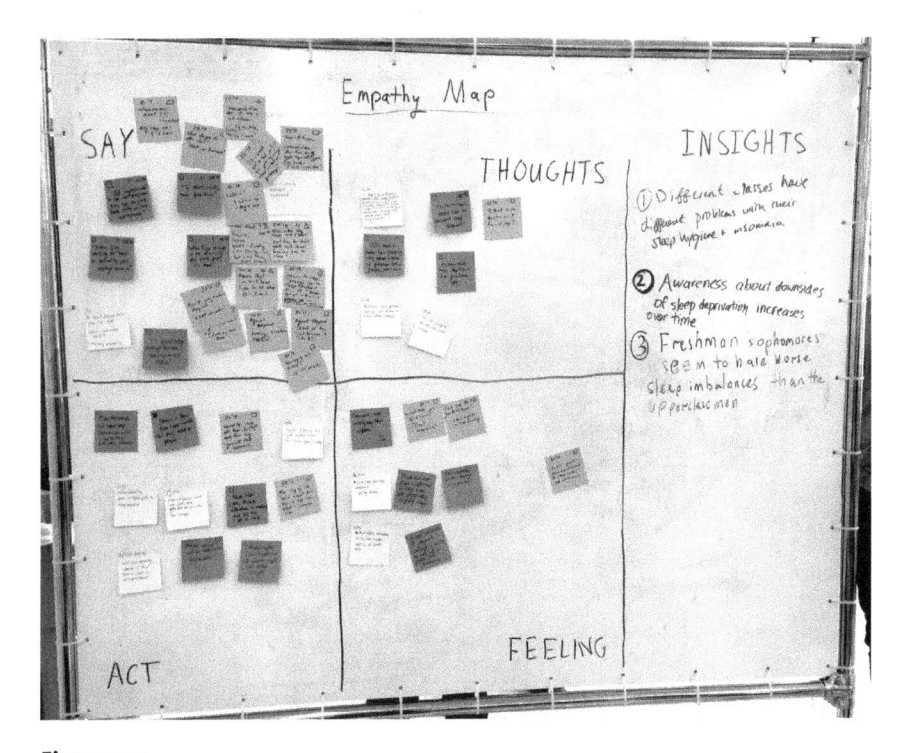

Figure 7.12
Empathy map to make sense of interview data. Princeton University students organized codes into four activities: say, thoughts, act, and feeling to help identify insights.

questions to guide the allocation of data to each activity. For example, Table 7.2 presents questions[20] that can help extract the necessary data.

While questions help add structure to the tool, empathy maps are heavily based on your and your team's interpretations of the data. Deciding in advance whether to use this tool helps you choose the appropriate data collection method to gather the right type of data to inform each activity's section. For example, observations and contextual interviews are useful methods to gather data about what participants "say" and "do", and help identify what they "feel".

To populate each activity's section, when possible, write down participants' actual words in response to the above questions on individual cards or sticky notes. Review the data, looking for participants' words describing

20 These questions are based on Bratsberg, H.M. (2012) *Empathy Maps of the FourSight Preferences*, Creative Studies Graduate Student Master's Projects, Paper 176, and Gray, D. (2017) *Updated Empathy Map Canvas* [online], Available at: https://medium.com/the-xplane-collection/updated-empathy-map-canvas-46df22df3c8a [Accessed 19 November 19].

Making sense of field data | 155

ACTIVITY	EXAMPLES OF QUESTIONS
See	• What did the participant see? • What did the participant notice in their immediate environment? • What were they watching or reading?
Say	• What did you hear the participant specifically saying? • Which expressions, phrases, or words did the participant use frequently?
Do	• What was the participant doing? • What activities did the participant enjoy? • What specific things did the participant do daily? • What behaviours did you notice?
Hear	• What did the participant hear from friends? • What sounds or words did the participant notice?
Think	• What would the participant actually be thinking? • Did their thoughts match their words and actions? • Were their thoughts manifested somehow through their body language?
Feel	• How did the participant feel? • How did the environment affect the participant's feelings? • How did the participant deal with their feelings? • Were their feelings manifested somehow through their body language or the tone/pace/volume of their voice?
Challenges	• What were the participant's fears? • What were the participant's frustrations? • Was there anything that was painful for them to do?
Gains	• What were the participant's dreams and hopes? • What was the participant trying to get done? • How did the participant measure success? • What results was the participant trying to achieve?

Table 7.2 Questions that can help extract information from participants to create an empathy map.

their view on the phenomenon or situation being studied and organize them in relation to each activity. Feelings and thoughts should be populated last, as they cannot be observed; you should infer them, based on what you learn from the other activities. Pay careful attention to clues that may communicate these activities, like body language, facial expressions, tone, or word choices. Write a summary of what you have learned about each person, focusing on explicitly describing inferred feelings and thoughts.

3. Interpret maps. After populating each activity of each template, each resulting empathy map represents a segment or group of participants that shares those similar feelings, goals, and needs. These similarities may or may not respond to their demographic characteristics. As an example, if you created three empathy maps, you would end up with three groups of intended audiences which could be developed further as personas. The next step is to create personas to represent each empathy map.

METHODS FOR SUPPORTING DATA INTERPRETATION

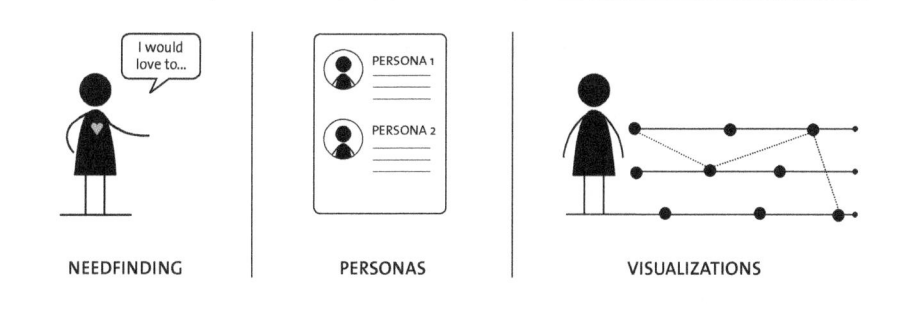

Figure 7.13
Three methods to support data interpretation.

Methods for supporting data interpretation

The three methods discussed in this section present guidance and support for interpreting coded data and generating meaning (Figure 7.13). In different ways, the goal of these methods is to reduce cognitive load and identify patterns by helping put your understanding and interpretation together in a new way. The resulting outputs represent the sense you made of the data, and are rough, often created by hand.

Needfinding

Purpose

The goal of *needfinding*[21] is to identify participants' unmet needs by interpreting and categorizing their words. This approach can be used to interpret coded data and infer needs directly from participants' actions and expressions or from contradictions between two actions—such as a disconnect between what they said and what they did. With this approach, you analyse coded data, looking only to identify unmet needs.

---

[21] McKim, R.H. (1980) *Experiences in Visual Thinking*, 2nd edn, Cengage Learning, and Patnaik, D. (2014) *Needfinding: Design Research and Planning*, Amazon. Needfinding is a qualitative research approach to studying people that can be used to identify their unmet needs.

How to use it

1. Look for verbs. Needs can be hard to identify. Read the coded data, paying special attention to verbs (needs), as these represent activities and desires with which people need help. Also, pay attention to contradictions between the different types of datasets collected (e.g. visual, audio, written). These contradictions often mark unrecognized or unarticulated needs, and problems that people have already developed ways to work around and don't realize that could be improved.

2. Create need statements[22]. Make a list of the identified verbs in any order and turn them into statements. Explain the potential need you think may be behind each verb, by adding many details, as shown in Table 7.3. When writing need statements:

- Use participants' actual words to describe each identified need.
- Use only positive phrasing.
- Be specific.
- Focus on the "whats" not the "hows" of a need. The "hows" indicate solutions because they are often nouns.

3. Organize need statements. Write or print each need statement on a separate card or sticky note. Group these according to the *similarity they express* (you can create an affinity diagram for this). Or organize sticky notes by *level of generality* (common, context, activity, and qualifier needs) (Chapter 1), in terms of whether you think a need may apply to every single person or just to that particular sample. After organizing all needs, visualize connections and hierarchies between needs, for example creating a mind map or diagram. Remember, your focus will be mostly on context and activity needs.

4. Prioritize needs. Rank needs based on what participants reported as high and low priorities or those pointed out by the majority. For example, your initial focus should be on the more frequently stressed needs. If you identify many needs but don't have a clear sense of which ones would be more important, conduct a short study, focused on establishing participants' level of importance for those needs.

Chapter 9 discusses how to translate identified needs into tangible actions to inform the information design process.

---

[22] Ulrich, K.T. & Eppinger, S.D. (2012) *Product Design and Development*, 5th edn, Irwin McGraw-Hill.

PARTICIPANT'S QUOTE	"WHAT"-FOCUSED NEED STATEMENT	"HOW"-FOCUSED NEED STATEMENT
"I'm so tired of getting to class late; it is so frustrating, I keep walking around trying to find the right building".	The person needs **to feel confident** that they know where they are going.	The person needs a map of campus.
"This is so frustrating, the app crashes whenever I don't use it for about two minutes, while I'm trying to figure it out".	The person needs **to understand** how to use the app.	The person needs a new app that doesn't crash after two minutes of not being used.
"The app didn't work very well, and this was annoying because I needed to transfer money online, even though it was Friday night".	The person needs **to be able** to access her bank account 24/7.	The person needs an app that works 24/7.

Table 7.3 Examples of need statements. To write need statements, focus on the "What" (verbs) not on the "How" (nouns), because nouns indicate solutions not needs.

Personas

Purpose

The goal of *personas*[23] is to provide a holistic view and understanding of different segments of a population or an intended audience, by moving away from abstract information towards the wants and needs of real people. A persona is a fictional profile within a targeted demographic, created with insights from research with the intended audiences, such as from thematic analysis or empathy maps. A persona doesn't show insights from only one participant. Each persona synthesizes data across many participants that share similar behaviours, needs, goals, and ways of thinking.

Personas are sometimes created purely from designers' imagination to summarize what an ideal user would look like. However, these personas may not reflect real users' needs. The more specific a persona, the more likely a design solution will work as intended, because specificity and precision help information designers make decisions to address concrete needs and likes. You can create personas at the beginning of the design process to segment the intended audience and gain a better understanding of the needs and expectations to consider in the development of a solution.

Detailed information presented in a persona often varies from project to project, but all personas include general dimensions of people's lives and habits—their goals, hopes, habits, likes and dislikes, where they perform their main activities (e.g. shop, spend time), what technology they use, who

---

23 Cooper, A. (1997) *The Inmates are Running the Asylum: Why High Tech Products Drive us Crazy and How to Restore the Sanity*, Indiana: Sams Publishing.

they interact with, what they eat, whether they are family-orientated, etc.—and project-specific dimensions—e.g. in an interface design project, a persona should include IT skill level, level of familiarity with devices, frequency of Internet use, spoken language, etc. Table 7.4 presents general and project-specific dimensions that can be included in a persona. Project-specific dimensions are often pre-determined during desk research and expanded later, after data coding and analysis. For example, to design a sports mobile personal app for a range of users, you might create personas based on research into people's sports habits (e.g. frequency, indoor or outdoor sports, type of sport) and their knowledge of different types of sports.

How to use it

1. Create template. If working alone, you can create templates in a Word document or notebook, but, if working with a team, use a bigger canvas, so other members can contribute to the creation of each persona. One common method is to use one flipchart sheet of paper as a template for each persona. List general dimensions that emerged from the analysis, and determine what project-specific dimensions will complement general dimensions. Then, divide each page, based on the number of general and project-specific dimensions you identified for each persona. Depending on how much information you have, persona templates can have different formats. Figure 7.14 shows three common ways of creating personas[24].

2. Segment intended audience. If you have created empathy maps, go to Step 3. If not, you first must get a sense of how many groups of users are within your intended audience:

- Read the coded data and look for shared needs and attitudes across participants, and any other relevant similar behaviours and stories.

- Extract and put together words and sentences exemplifying groups of participants with shared needs, behaviours, struggles, and attitudes. Write these words and excerpts on individual sticky notes or cards.

3. Populate template. Populate each dimension with representative words and excerpts from the data. Write directly on the sheet or use the sticky notes you wrote in the previous step to stick on the corresponding section. In either case, attach a colour to each dimension and use it consistently across all personas.

---

[24] Based on *Personas* article [online], Available at https://www.usability.gov/how-to-and-tools/methods/personas.html [Accessed 10 January 2018].

DIMENSION	DESCRIPTION	EXAMPLES
GENERAL DIMENSIONS		
Representative name	• What phrase best represents this person's needs and behaviours?	"Paul, the energized commuter"
Representative quote	• What did the person say that most accurately represent their need? Or key characteristics?	"I love waking up early and getting ready for work. I don't mind commuting during rush hour, that wakes me up and gives me energy for the rest of the day"
Demographics	• Who is the person?	Name, gender, age, culture, income, location, marital status, etc.
Behaviour	• What behavioural patterns did you notice? Behaviour over time-day, week, month, year.	Moods, routines, usages, daily use patterns, social interactions, etc.
Personality	• What is their attitude?	Introvert/extrovert, curiosity, creativity, cost sensitive, etc.
Likes / Dislikes	• What do they like and dislike doing? • Why?	Likes waking up early and drinking coffee with sugar, doesn't like walking.
Social environment	• What are their social interactions?	Family, friends, sisters, mostly co-workers.
Skills and knowledge	• What are their areas of expertise? • Where do their excel?	High IT skills, good with numbers, great at sports, poor cooking skills, etc.
PROJECT-SPECIFIC DIMENSIONS		
Feelings	• What do they want to gain from the experience of using the design?	A feeling of... (freedom, reassurance, confidence, etc.)
End goals	• What do they want to accomplish by using the design?	Get to work faster, save money, be successful, etc.
Skills and knowledge	• How proficient are they in this domain? • How much do they know about the topic explored or design tested?	Very proficient, little confident, knows the basics to know how to use the design, needs direction, etc.
Pain points	• What do they struggle with? • What things are they lacking?	Challenging situations, uses, etc.
Environment	• Where did they express the lack of something? • What environmental factors are relevant for them in the use of the design?	Uses design in car, on public transportation, outside, inside, on a plane, etc.

Table 7.4 Key general and project-specific dimensions that a persona can include.

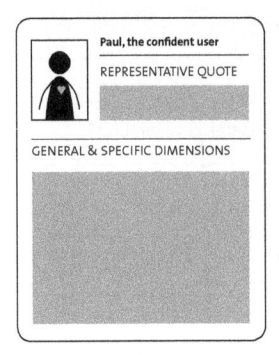

Detailed explanation of each dimension **presented as a story**. One-page template for each persona.

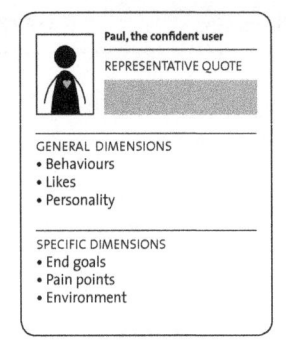

One-page template for each persona **only showing key dimensions** related to the project.

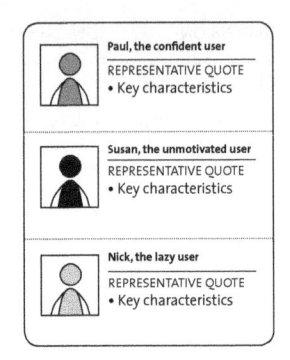

One-page template **presenting three or more personas and a short description** of their key characteristics.

Figure 7.14
Formats for creating personas.

4. Build empathy. Once all dimensions are filled, summarize extracted words and sentences for each persona, including:

- Personal information: name, age, where they live, hobbies and interests, likes and dislikes, needs, behavioural patterns.

- Professional information: occupation/affiliation, level of experience, skills.

- Technical information (if relevant).

- Project-specific information.

To increase the degree of empathy and facilitate the distinction between each persona, use participants' actual words and quotes for authenticity. Add any visual material or text (e.g. narrative, table) that helps clarify the essence of each persona, as well as a picture or drawing to each persona, showing how a representative person of that segment would look, based on data presented.

There is no fixed number of personas you should create, but having more than four different personas may indicate that some aren't distinct enough and could be combined. At this stage, personas are working documents, helping you figure things out and interpret the data[25]. It is ok if they look rough and not very polished. These same personas can then evolve and be used to report findings to clients. In this case, they are cleaned up and designed in higher quality as described in Chapter 8.

---

25 Weinstein, L. (2015) *An Inside Look at Design Tools in Development Work: User Personas in Context* [online], Available at: https://reboot.org/2015/06/18/user-personas-in-context/ [Accessed 31 August 2017].

Visualizations

Purpose

As in previous steps in the research process, visualizations[26] can also be created to support sensemaking. Their goal is to provide structure and stimulate thinking, without prescribing definitive paths or causing distraction. The challenge of visualizing qualitative data is creating graphics that keep the power of the text, its meanings, and emotions[27], without oversimplifying or misrepresenting the data. At this stage, visualizations should support your process not make it harder. These visualizations don't have to look "pretty" and there's no need to use fancy software to create them. They are just to help make sense of the data. These visualizations aren't infographics. This is an important distinction from visualizations created for sharing findings with clients (Chapter 8).

Many types of visualizations can be created at this point. Some transform qualitative data into quantifiable units (e.g. bar charts, line graphs, scatter plots)[28]. To create these, data should be coded and broken down into component parts, such as individual words, sentences, themes, or narratives[29], that you then count, categorize, or connect with other components[30]. For example, these types of visualizations can support visual content analysis. Other types, the focus of this section, represent the "why" and "how" of the data; they don't quantify the data. Here, an overview of these visualizations and how to use them to make sense of qualitative data is provided.

Portraits

These visualizations represent the "whats" (e.g. artefacts, contexts, tools) and "whos" (e.g. participants, personas, other parties) of a study. They need to be simple, so they don't distract from the essential meaning of the concept, but they also need to be representative to help distinguish and connect data with concepts[31]. For example, if working with various types of users, you can create a portrait to help distinguish each user type. In this case, each portrait could

---

26 Miles, M.B., Huberman, A.M., & Saldaña, J. (2013) *Qualitative Data Analysis*, SAGE; Henderson, S. & Segal, E.H. (2013) Visualizing qualitative data in evaluation research, *New Directions for Evaluation*, 2013(139), 53–71.
27 Bernard, H. & Ryan, G. (2010) *Analyzing Qualitative Data: Systematic Approaches*, Thousand Oaks, CA: SAGE.
28 Patton, M.Q. (2002) *Qualitative Research & Evaluation Methods*, 3rd edn, Thousand Oaks, CA: SAGE.
29 Bernard, H. & Ryan, G. (2010) *Analyzing Qualitative Data: Systematic Approaches*, Thousand Oaks, CA: SAGE.
30 Content analysis works with a similar approach: quantifying qualitative data.
31 Roam, D. (2013) *The Back of the Napkin: Solving Problems and Selling Ideas with Pictures*, London: Portfolio.

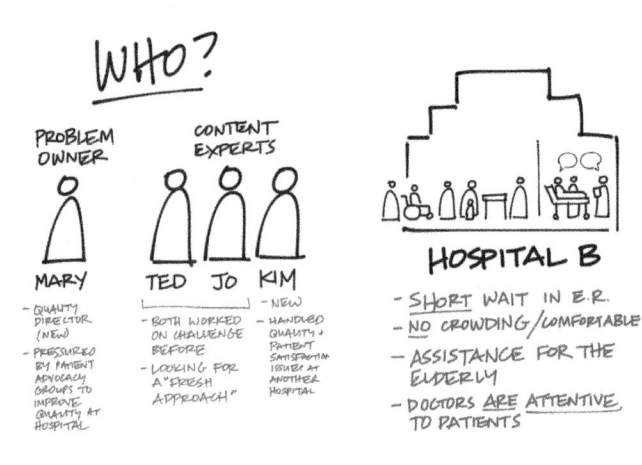

Figure 7.15

Representations of concepts to visualize findings.

show a distinctive aspect of each user. Regardless of what you create, anchor the visualization's meaning by attaching a word. There are two large groups of portraits, based on what they represent: pictographs and ideographs.

Pictographs. These are visuals created to represent something or someone described in the data, such as objects or individuals. They don't have to be realistic, but they need to resemble what they represent; they can be abstract if that helps visually distinguish key data.

Ideographs. These are visuals created to represent a concept, idea, or theme. They aren't literal representations; they can be abstract.

For example, adding a visual to represent concepts or key characteristics helps the team remember the essence of the intended audience or a main finding. Adding visuals to represent data also helps humanize findings, as shown in Figure 7.15. Rather than representing data insights in an abstract form only using words, you can create a visual system that represents the essence of each word.

Connections and relationships

These visualizations help represent connections, meaning, patterns, and themes that emerge within a participant, across participants, or across various datasets. Connections can be visualized in many ways:

Diagrams. These visuals deconstruct textual data into a visual (graphic or typographical) structure that represents connections between individual units or components, as in mind maps. Create diagrams to visualize spatial and directional relationships, to show patterns at a point in time or how

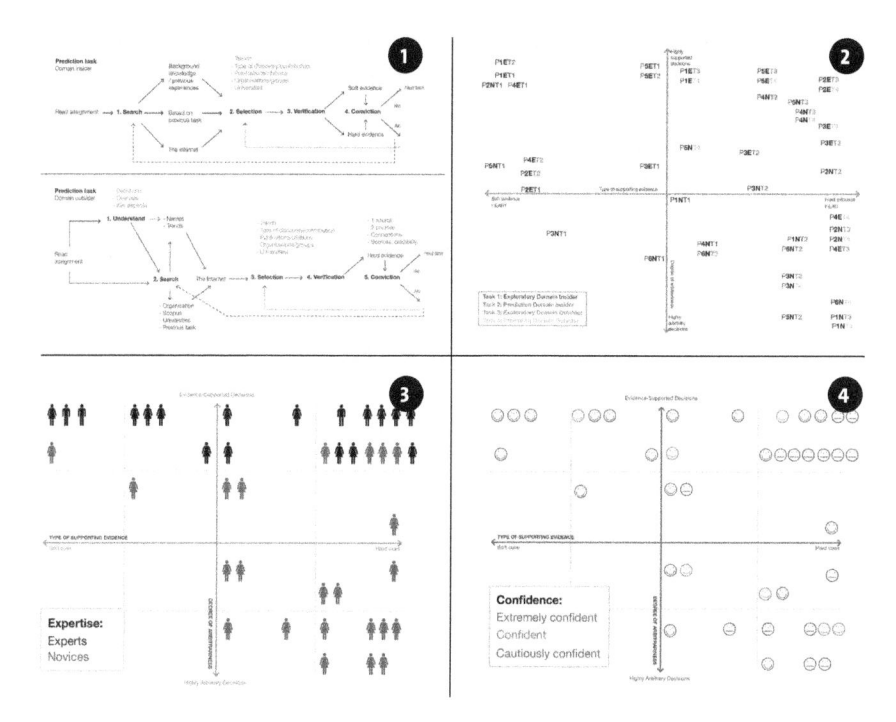

Figure 7.16
Evolution of diagrams for visualizing connections and findings.

things are organized or interconnected. Also, diagrams help show connections and hierarchical structures of concepts or themes, ranked in grades, levels, or classes, as shown in Figure 7.16.

Flowcharts. This is a type of diagram that shows a sequence of decisions and steps involved in a workflow or needed to perform a process. Steps are visualized as boxes of different shapes (e.g. diamonds, squares, circles) and ordered through lines and arrows symbolizing connections. Unlike other diagrams, flowcharts are schematic and often text-based.

Evolution and sequences

Use these types of visualizations to present changes over time; they can take many forms:

Timelines. These visuals display data in chronological order, making connections more apparent, or can be used to better understand how something unfolds over time. Timelines can be text-based or combine text and images.

Making sense of field data | 165

For example, you can create a timeline to visualize key common events that occur during a work day reported by participants.

Process. This type of diagram visualizes a sequence of organized steps, performed to achieve an end; you can't change the order in which steps are presented, and a temporal component is often attached to each step. Unlike flowcharts, process diagrams often include pictographs or ideographs to illustrate each step. Process diagrams can help reveal the steps, actions, and time frames involved in an activity that are often invisible to participants, clients, or other stakeholders. Create a process diagram to show how to do things, how something works, or to illustrate specific steps, gaps, or areas for improvement. For example, you can visualize the sequence of steps that a person needs to book a flight online or use a train ticket vending machine.

Journeys. These visuals represent an experience of how a persona achieves a particular objective in the form of connected sequence tasks. In contrast to process diagrams, journeys represent the events or places a persona goes through, without a goal in mind. Journeys show:

- **Touchpoints:** instances where a persona interacts with other actors or uses a particular design to achieve their objectives
- **Pain points:** challenges or struggles a persona faces along the way
- **Magic points:** parts of an experience or design that work well for a persona

You can create a journey to represent how to go from A to B, the experience of using a service, getting through the day, or what a person needs to get vaccinated. For example, a journey can cover all the steps, from when a persona first becomes aware of a design until they stop interacting with it, showing the different points of contact and their feelings.

As shown in Figure 7.17, to create a journey:

1. Identify a persona's points of view, experiences, and emotions about the journey's main subject.

2. Identify and extract broad steps that a persona goes through to achieve their goal. Use these to create the journey's general structure.

3. Identify activities and plot them chronologically as a sequence, following the general structure you created.

4. Simultaneously, identify people involved in the journey and at each step, and any touchpoints, pain points, and magic points.

5. After mapping all this data as a sequence (this could involve parallel tracks), identify any gaps and unknowns. These are parts of the

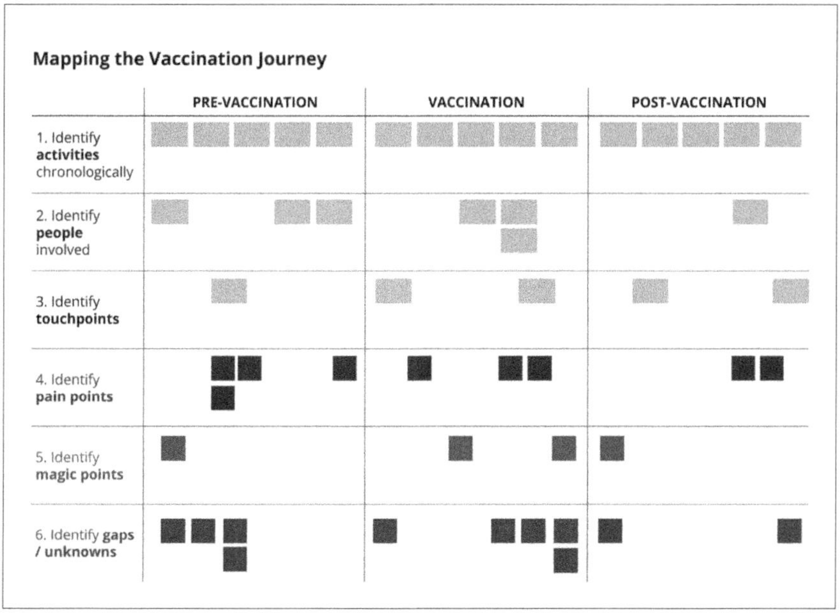

	PRE-VACCINATION	VACCINATION	POST-VACCINATION
1. Identify **activities** chronologically	▪▪▪▪	▪▪▪▪	▪▪▪▪
2. Identify **people** involved	▪ ▪▪	▪▪ ▪	▪
3. Identify **touchpoints**	▪	▪ ▪	▪ ▪
4. Identify **pain points**	▪▪ ▪ ▪	▪ ▪▪	▪▪
5. Identify **magic points**	▪	▪ ▪ ▪	
6. Identify **gaps / unknowns**	▪▪▪ ▪	▪ ▪▪▪ ▪	▪

Mapping the Vaccination Journey

Figure 7.17
Template to create a doctor and patient vaccination journey. The first column presents six key steps to extract data, and the following columns show three main stages in the journey. The squares represent post-its which can be colour-coded as needed.

journey that are unclear and parts where you would like to learn more and explore further.

6. You can add images or visuals to personalize the journey and facilitate empathic engagement.

Figure 7.18 presents an example of a vaccination journey created from research data following these steps.

Process diagrams and journeys should be created in line with the study's goal and using the data gathered, for example, to learn how something is done (exploratory) or to gain a better understanding of how a current or new design is used (evaluation). In both cases, they help better understand the impact of a problem, participants' experiences, the use of a specific design, or the component parts of an experience. In the case of journeys, visualizing touch, pain, and magic points helps see which parts of an experience or design work well (magic points) and which might need improving (pain points).

The next chapter discusses the use of these visualizations and the creation of a story to report findings.

Making sense of field data | 167

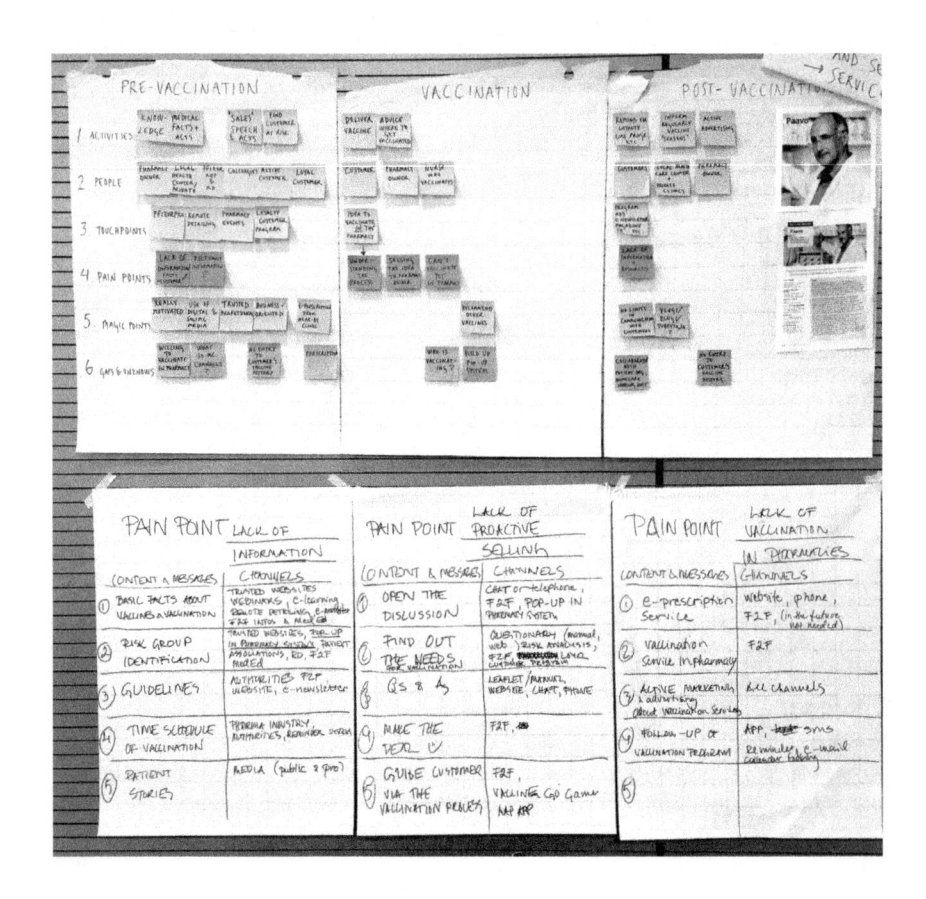

Figure 7.18
Vaccination journey created using the template from Figure 7.17.

PART III:
Communicating findings

8 Reporting field research findings

After making sense of data, the next step is communicating what you learned and how those learnings inform the information design process so the team can make subsequent decisions supported by evidence. Information designers often create visual stories, storyboards, and visualizations to share study findings and learnings[1]. Cleaner versions of most of the visualizations discussed in the previous chapter frequently make their way into final presentations and reports. This chapter discusses these and other ways you can use to communicate findings.

Dimensions for communicating findings

While the communication of findings comes at the end of the research process, it is important to think about it at the beginning. The following three dimensions determine the tone, style, length, and content of how findings will be shared:

- **When:** Design process stage
- **Who:** Audience
- **What:** Format

Design process stage

Depending on where you are in the process, the audience (the rest of the team, intended audience, or clients) and format to communicate findings vary, because findings will be used in different ways. In the early stages, insights are communicated to give focus and initial direction to the development of ideas; in the middle, learnings validate a direction and are often shared in the form of recommendations to move forward from conceptual ideas to prototypes or convince a client on the appropriateness of an idea.

---

1 Sleeswijk Visser, F., Van der Lugt, R., & Stappers, P.J. (2004) The personal cardset—a designer-centred tool for sharing insights from user studies, in *Proceedings of Second International Conference on Appliance Design*, Bristol, 157–158.

Later in the process, findings help optimize a design and are used to inform how to make improvements to a prototype or to act as stepping stones for further research.

Audience

Findings can be communicated internally, to your team, or externally, to intended audiences (if taking a participatory approach), clients, or other stakeholders. In the former scenario, the focus will be mostly on *what* you learned and how those learnings can inform the information design process[2], while, in the latter, the focus will expand and also include *how* you did what you did, so they can follow your trail. In either case, while you do have to demonstrate the quality and credibility of the findings when reporting the study to "justify" why you did what you did, the focus is less on the methodology and more on the learnings and their applicability. Clients want to see concrete evidence, indicating that a design idea would be successful and your design decisions are the right ones, while your team wants to see how findings could inform the rest of the development process.

Format

You may think that field research should be reported in a hefty tome, mostly using textual narrative, and full of academic jargon. However, neither of these characteristics is necessarily true. Field research can also be communicated in a brief and concise way, employing plenty of visuals, and plain, clear language. Formats vary depending on where in the information design process you are, the audience, and context in which findings will be used. For example, you can use more interactive and design-driven techniques, focused on presenting findings and how to move forward in the information design process, to communicate with the rest of the team. These techniques include visualizations, storyboards, personas and scenarios, posters, or video. Or you can use more conventional sequential narratives including polished digital visuals, as in reports and shown in Figure 8.1[3], executive summaries, or presentations, created as PDFs, Keynote, or PowerPoint stories, when reporting to clients.

---

2 Patton, M.Q. (2002) *Qualitative Research & Evaluation Methods*, 3rd edn, Thousand Oaks, CA: SAGE.
3 Pontis, S. & Blandford, A. (2015) Understanding "influence": An exploratory study of academics' process of knowledge construction through iterative and interactive information seeking. *Journal of the Association for Information Science and Technology*, 66(8), 1576–1593.

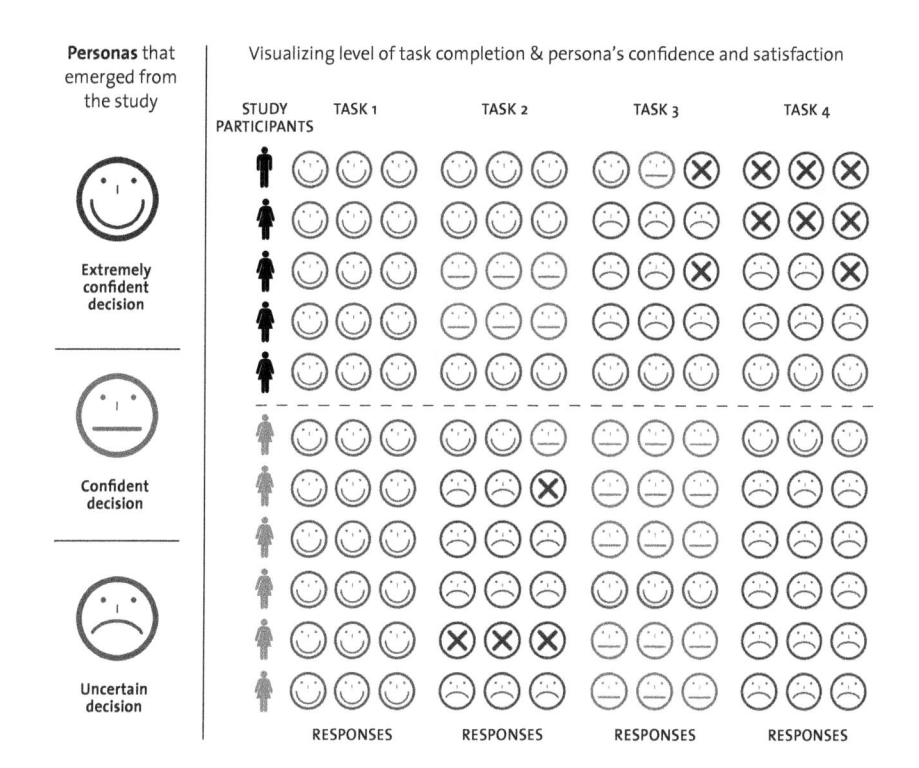

Figure 8.1
Visual table to share study findings. From this study, conducted at University College London, three personas emerged which are indicated on the left column. Participants' level of confidence with their responses was visualized as a table to make connections between the type of data used to make decisions and whether tasks were successfully completed clearer.

Creating authentic stories

Above all, a field study is a **story** and should be reported as such. Creating a compelling story helps articulate findings more clearly. In addition, having the study explained (in some form) makes it accessible for future reference and to inform other projects.

When reporting a field study, the goal of your story should be twofold: ensure that what you communicate comes across as valid and credible (addressing validity and quality criteria; Chapter 3), and is genuinely reflecting participants' experiences. It is therefore essential to have a clear focus and be engaging, not necessarily including everything you did and learned. Remember: your audience is people. To show authenticity, the story should transport them to the context (time and place) where the study took place[4]. Regardless of the format, connect abstract insights and learnings to something concrete that the rest of the team or your clients can relate to and use to empathize with participants. Two ways to help your audience build empathy with a study's participants and better understand their lives and experiences[5] follow:

- **Visual material:** Field research data is rich but also contains intangible aspects that can be best communicated through visualizations, such as portraits, photos, diagrams, journeys, drawings, videos, collages. Participants should be shown as real people not just as abstract numbers or acronyms.

- **Unfiltered descriptions:** Raw data gathered from the field in the form of participants' actual words helps participants to be seen as real people, supports any claims or findings you are sharing, and provides colour and details about participants' ways of seeing the world and their everyday lives.

Authentic stories put findings in context and help bring data to life, by providing concrete examples, details, and colours, and to convey participants' experiences but also any interactions with tested prototypes. Using visual material and unfiltered descriptions in a story helps create a compelling narrative.

You can create a story that provides an overview of the **whole study** or one that focuses on a **specific moment** or **key findings of the study**. The following sections provide a story structure to share a complete study and discuss various formats in which you can communicate specific parts of a study, depending on its use.

How to share the whole study

Formats frequently used to share the whole study are **reports** or **presentations**, although you can also use these formats to only communicate a part of the study. While a story's format and content can vary, the overall structure will mostly remain the same with some variations[6]: an overview or summary of the study, followed by the main storyline, closing with conclusions, recommendations, and further steps. The following are pointers to help organize and present the study's research process steps and what you learned in a clear and compelling way.

---

4 Chipchase, J. (2017) *The Field Study Handbook*, 2nd edn, Field Institute.
5 Sleeswijk Visser, F. (2009) *Bringing the everyday life of people into design*, PhD thesis, Delft University.
6 Roam, D. (2014) *Show and Tell: How Everybody Can Make Extraordinary Presentations*, London: Penguin.

Study overview

The beginning of the story succinctly and clearly explains the study's goal, the research strategy, and the methods used to gather and analyse data. It also includes:

- **A review of similar cases.** This section puts your study in context and demonstrates that you investigated the strengths and weaknesses of prior related work. Furthermore, it backs up some of the decisions made throughout the study. Include a short description of similar existing cases that you used or read to inform the study.

- **A summary of findings.** For most stakeholders, this is the story's core. Avoid using very technical or unnecessarily complex words, even if delivering an internal presentation to your team. Be concise and unpack each finding. Provide context and explicitly stress the connection of each finding to the study's initial research questions.

Main storyline

This is the bulk of the story. Here, you provide the study's data, evidence, learnings, and other relevant supporting details. One way to structure this content is to use the same research questions, determined at the beginning of the study (Chapter 4), or the themes that emerged from the analysis (Chapter 7), as section headers. For example, answers to each question or theme can be used as a header or section divider in the storyline. This part of the story includes:

- **Themes and findings.** Introduce emerging themes and link them together as the narrative of the story, and use visualizations to create 'a logical chain of evidence'[7] to support findings and emerging ideas. Make the story more robust by using related insights from your initial desk research (e.g. client meetings, reports, websites, previous cases, etc.) to support themes.

- **Descriptions and evidence.** Add a human layer and authenticity to the story, by explaining each theme, finding, or idea, using participants' actual words and expressions, and using excerpts to provide more illustrative descriptions. The key is presenting a balanced proportion between *description* (what participants said) and *interpretation* (identified themes and findings). Descriptions are the selection of multiple, illustrative excerpts and actual words from the data, used to support findings.

---

[7] Miles, M.B., Huberman, A.M., & Saldaña, J. (2013) *Qualitative Data Analysis*, SAGE.

Interpretation is an explanation of your thought process and analysis that helped you generate meaning and identify what you are reporting. Combine descriptions with your interpretation in narrative form, explaining what the excerpts mean in the context of the design project. Include quotes in participants' own words, to illustrate their nuances, complexity, and depth of thought, and to help readers understand different degrees of significance of findings. For example, clarify when patterns are clear and strong, and when they are just insinuated, and include participants' testimonies that explain *why* those patterns seem to be strong and others weak.

- **Validation of findings.** This additional evidence helps instil confidence in the reported findings. Here, you explain any triangulation techniques you used (Chapter 3). Mostly, you will have conducted the field study solo or with another person (unless you work in a large information design company where there is a big team involved in the study); reports frequently reflect this unilateral approach, as they are heavy on the presentation of findings and descriptions but light on how those findings were reached. To make findings more robust and credible, provide a short but detailed step-by-step description of the analysis process. One way is to use a diagram, showing the sequence of steps followed during data analysis.

It is also especially important to be open, honest, and transparent when communicating field research findings. If something didn't go as planned, insights contradict current work, or other, contradictory, views presented by the data were encountered, then point these out, and explain any limitations of the study. Furthermore, include the criteria used to recruit participants and clarify how you generated meaning from the data.

Conclusions, recommendations, and further steps

The end of the story focuses on explicitly connecting the study with the overarching information design project. Further steps should clearly indicate the applicability of findings, and can also suggest other studies that could be conducted to validate findings.

- **Application of findings [Chapter 9].** Presenting the relevance, utility, and applicability of findings as actionable items rather than abstract or generic statements is particularly important in this type of report. Anchor each finding with a relevant need and/or characteristic of the project, and communicate them in a way that this is actionable for the

TOOLS FOR SHARING KEY PARTS OF THE STUDY

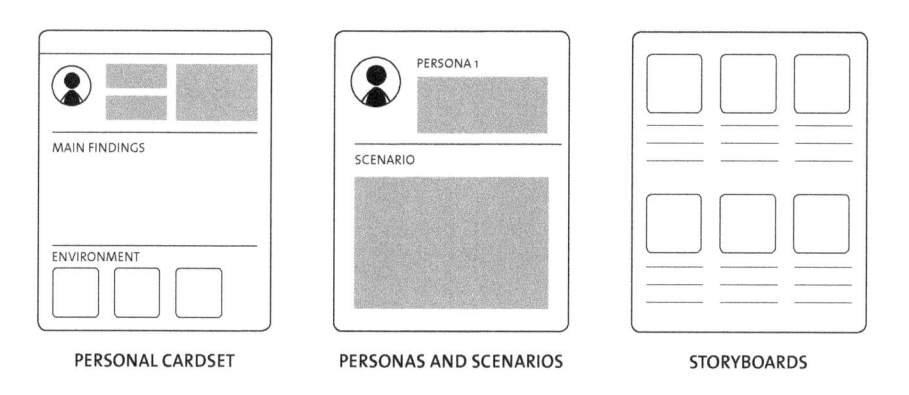

Figure 8.2
Tools to share key parts of the study internally with the team or externally with clients and intended audiences.

rest of the design team[8]. For example, findings can take the form of design recommendations, principles, guidelines, or a list of changes that the team should use to move forward in the design process. Stress the clear link between each finding and how it informs the design process, and explain the direct link between findings and further steps. These links can also be visually explained with sketches or visualizations that communicate each need and finding and show how they connect back to the project and the information design process.

How to share key parts of the study

In some cases, rather than explaining the whole study, you will only share key findings with the rest of the team, clients, or intended audiences for active collaboration. This section discusses three tools, as shown in Figure 8.2, that integrate findings and context placing people at the centre, and that you can use to promote discussion, gather inspiration, generate ideas, tell stories, and build understanding. You may have used some of these tools in earlier steps of the research process for different purposes (Chapters 5 and 7).

---

8 Beyer, H. & Holtzblatt, K. (1998) *Contextual Design: Defining Customer-Centered Systems*, San Francisco: Elsevier, and Holtzblatt, K. & Beyer, H. (2014) Contextual Design: Evolved, *Synthesis Lectures on Human-Centered Informatics*, 7(4), 1–91.

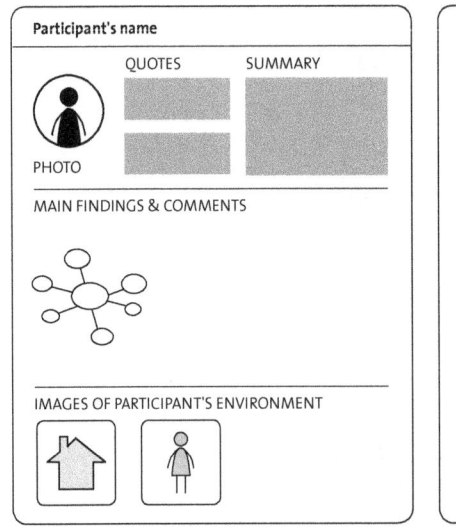

PERSONAL CARDSET TEMPLATE

FRONT

BACK

Figure 8.3
Template to create a personal cardset.

Personal cardsets as working tools

If your study involves a small sample (fewer than ten participants), you could create *personal cardsets*[9] (Figure 8.3). In contrast to personas, each card will show one participant's data, with the set representing your entire sample. The front of each card will show each participant's personal information and raw data, and the back your interpretation of that data and findings from the analysis. Personal cardsets are often created as a working tool for designers to share research findings with the rest of the team during the early stage of the design process. Having access to all participants' raw data and findings in a structured way can help the team gain a deeper understanding of intended audiences and enhance idea generation.

- **What to include:** In **the front** of each card, write a summary of each participant, including a title, one or two quotes representing their per-

[9] Sleeswijk Visser, F., Van der Lugt, R., & Stappers, P.J. (2004) The personal cardset—a designer-centred tool for sharing insights from user studies, in *Proceedings of Second International Conference on Appliance Design*, Bristol, 157–158.

sonality, their picture and name; to maintain confidentiality, the latter two shouldn't be real, but they should be related to their actual characteristics. For example, if a participant is a blond curly-haired child, the picture should also have these features. You can also include images of environments or objects relevant to the participant. In addition, include your interpretations of the data, emerging themes across participants and any interesting findings that emerged during data analysis, in diagrammatic form. In **the back** of each card, include a transcript of the participant's data, gathered during the study. If your dataset is too large, only include important parts of the transcript.

- **How to do it:** For each participant, create a letter or A4 size card made of cardboard or foam board. As a working tool, you can also laminate both sides of the cards so team members can write notes on them and indicate connections between cards without messing them up. Depending on the size of the study, and the size of the team, you can create bigger size cards to also allow more varied forms of contributions from the team (e.g. drawings, sketches, diagrams).

Personas and scenarios as storytelling tools

Chapter 7 discussed the creation of personas as a way to identify intended audiences and users, and synthesize their needs. Personas can also be used as a storytelling tool to share findings with clients and initiate conversation. A way to help clients empathize with personas is creating a *scenario*. This is a hypothetical story, including great level of detail based on insights gathered from the study that describes how a persona would behave or act in a specific situation relevant to the design project. This situation could represent an interaction with a specific design or describe a routine indicating tasks that could be better supported with a new design or redesign of an existing solution.

- **What to include:** If you haven't created personas during data analysis, return to your analysed data and do so before you start writing a scenario. If you have, select one persona that you want to work with first, this can be the most intriguing, unexpected, or representative of your intended audience. At the beginning of the scenario, indicate who the persona is, why the persona is in the situation you are about to describe (e.g. complete a task, interact with a tool), and what their goals are.

- **How to do it:** Scenarios are often text-based or a series of written bullet points, but they can also take the form of a flow chart. Describe the steps or tasks the persona needs to perform to successfully achieve a goal or fail to do so. Give enough detail to help the reader construct the sequence of events necessary for the persona to address the goal and understand the challenges that each step involves.

In contrast to the use of personas during data analysis, in this case, personas and scenarios should be polished and well presented.

Storyboards as visual storytelling tools

In contrast to the use of storyboards discussed in Chapter 5, at this step in the research process, storyboards can be used to help articulate personas' needs, illustrate scenarios, or share stories you learned from the study. Each moment, task, or step described in a scenario is represented by drawings, sketches, or collages, and put together as frames in a narrative form, as shown in Figure 8.4. Storyboards can represent many storylines, based on different personas, or focus on one specific scenario, representative of relevant behaviour or experience. The storyboard's goal is to show a persona's story, by describing:

- Moments in which they may need support or help to accomplish a goal within a given scenario or complete a specific task. For example, these can be used to illustrate opportunities for ideation during the early stage in the design process when you still don't have a prototype or concept.

- Interactions between them and the prototype that helped them accomplish their goal. For example, these can be used to illustrate magic points, touchpoints, and pain points during middle- or late stage in the design process, when testing a concept or prototype, and determine which areas are working well and which ones should be improved.

Each storyboard should be standalone and clearly communicate a scenario:

- **What to include:** Return to your analysed data and created personas, identifying which part of the story or scenario you want to tell. Also, determine the use of the storyboard. If you want to show a persona's needs, ensure you select a part of the story where needs are manifested. If you want to show a persona's interactions with a prototype, choose various moments to reflect diversity.

- **What to show first:** Determine the order in which to present the interactions and whether any of the frames is more important.

- **How to do it:** If you are creating a storyboard to discuss learnings with the team, it can be rough, but pay attention to the style you use and the details you include in each frame, as each communicates something about what you learned and the persona's lifestyle. But if you are creating a storyboard as a tool to start conversation with clients or intended audiences, it has to be of higher quality and polished design.

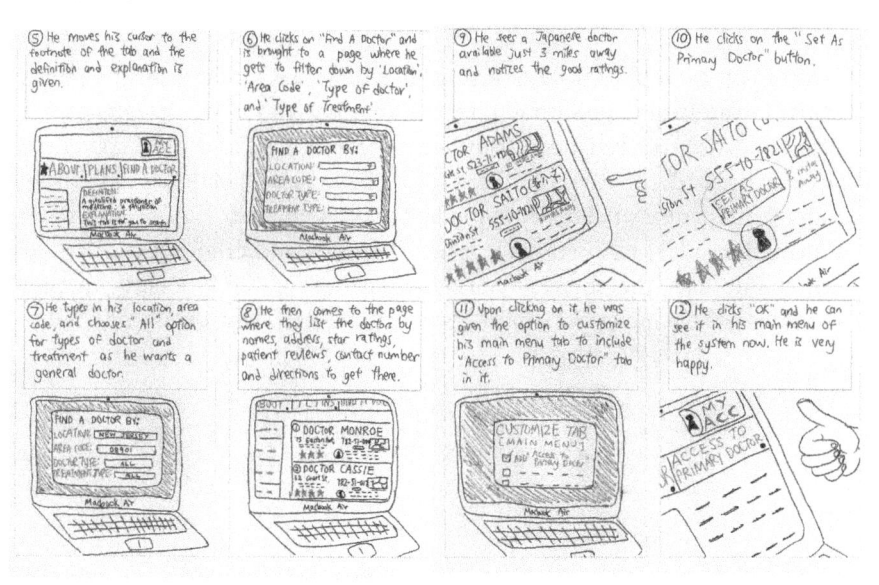

Figure 8.4

Storyboards from contextual inquiry sessions. After analysing data, Rutgers University students created storyboards to illustrate pain points in the use of healthcare websites.

Reporting field research findings | 181

- **Stress interactions and gaps:** Emphasize the persona's needs or interactions with a prototype, by adding colour or changing the style or line weight.

The same richness you capture during fieldwork should be communicated by adding details into each frame of the storyboard. In some cases, the use of *photoboarding*[10] may communicate a more accurate portrait of a persona, or photomontage may be enough to convey the sense of a specific space. In other cases, hand-drawn visual material may give the rough sense reported by participants. It is important that the visual style represents the personality or state of mind of the given persona. Frames can also include short descriptions to anchor the meaning of visual material.

Unlike reports, these three tools can spark discussion in a meeting, gather feedback from users, determine where a prototype needs improvements, and easily outline how to make them. But all of them, reports, personal card-sets, personas, scenarios, and storyboards help bridge theory and practice, by showing clients how research findings could be used in the context of the information design process, and by helping the rest of the team start a more focused discussion on the application of findings and definition of action items.

After you have put together a story or shared key findings of the study, return to the study's goal to determine whether you have answers to the initial questions or you need to collect more data. Ask yourself: Do these findings help address the initial goal? Do I have new questions? Can I fully understand people's needs? Are any behaviours unclear? Do I need further information such as from other intended users or a different setting?

If you do have the answers you were looking for, the following chapter discusses how to move from findings to actionable design decisions.

---

10 Photoboarding is a technique for creating storyboards that combines role-play and photography. Based on a persona's description, designers and other members of the team act out a scenario that will be included in the storyboard, while another team member takes photos. These are then printed and manually edited (e.g. detail added with pencils and markers, and parts removed); the final result becomes the visual material for each frame. (Sleeswijk Visser, F. (2009) *Bringing the everyday life of people into design*, PhD thesis, Delft University.)

9 Bridging to design: from findings to actionable design decisions

We have come full circle: Part I introduced the information design process, indicating the need for research to strengthen some of the steps so that designers can make more informed design decisions and create solutions more aligned to the intended audience's needs. Part II dived into the research process, providing guidance to conduct a field study. After data analysis, the next greatest challenge is using research findings as part of the information design process. In many cases, rich and high-quality data is gathered and analysed, and findings are communicated to clients, but then the design team struggles to use these findings to advance the development of a design[1], as illustrated in Figure 9.1. This challenge relates to the intangible nature of field research findings, often reported as feelings, behaviours, experiences, or motivations.

Figure 9.1
The challenge of turning findings into actionable items. After a design team completes a field study and communicates the main findings to clients, the team may struggle to apply the learnings to evolve a concept or revise a design.

1 Beyer, H. & Holtzblatt, K. (1998) *Contextual Design: Defining Customer-Centered Systems*, San Francisco: Elsevier, and Holtzblatt, K. & Beyer, H. (2014) Contextual Design: Evolved, *Synthesis Lectures on Human-Centered Informatics*, 7(4), 1–91.

This chapter returns to the information design process and focuses on presenting techniques to translate findings and insights into actionable items, so that they can be used to inform the design process and create a new design from data. More experienced information design researchers bridge findings and design in an intuitive way. Steps and techniques discussed in the following sections externalize the thinking process needed to create that bridge and present it as a structured way to provide guidance. These techniques are indicative but not prescriptive; they represent just ways of managing this part of the process.

Understand findings

Often, findings are communicated as recommendations, guidelines, or changes for the design team to implement. In the best case, these could be a list of features or dimensions to start developing ideas or to improve an existing design idea, or a list of problems that should be addressed in a prototype. But in many cases, findings aren't so tactical or clearly articulated as actionable tasks. If this is your case, the first step is to thoroughly understand what you learned from the study; that is, your findings. Then, findings need to be organized and interpreted; after this make them tangible, so that they can be converted into design concepts, and actionable items. Figure 9.2 summarizes this process.

Occasionally, a finding manifests itself as an explicit, tactical actionable item, such us: "having bigger squares rather than circles is better for us to understand how to complete the form". Or a pain point is frequently reported: "I always struggle to find the submit button"; this is likely to indicate that it is of great importance for the audience and, therefore, could be a high priority for your team to address first. However, most likely, participants won't know that something doesn't work for them; while some may be able to pinpoint it, they won't be able to express how to fix it (even if they do articulate this, their design suggestions should be taken with a grain of salt and interpreted).

This is why it is important that you take the time to understand what each finding is really about, what it may be implying or indicating. Consider that:

- each finding should be seen and understood in the context of the project, and not as an isolated entity.
- each finding will be used in a different way, based on the study's goal and where you are in the information design process (early, middle, late).

A common mistake is to interpret findings as direct solutions for the design problem at hand. However, a finding doesn't indicate a solution; it is an

HOW TO TURN FINDINGS INTO ACTIONABLE DESIGN DECISIONS

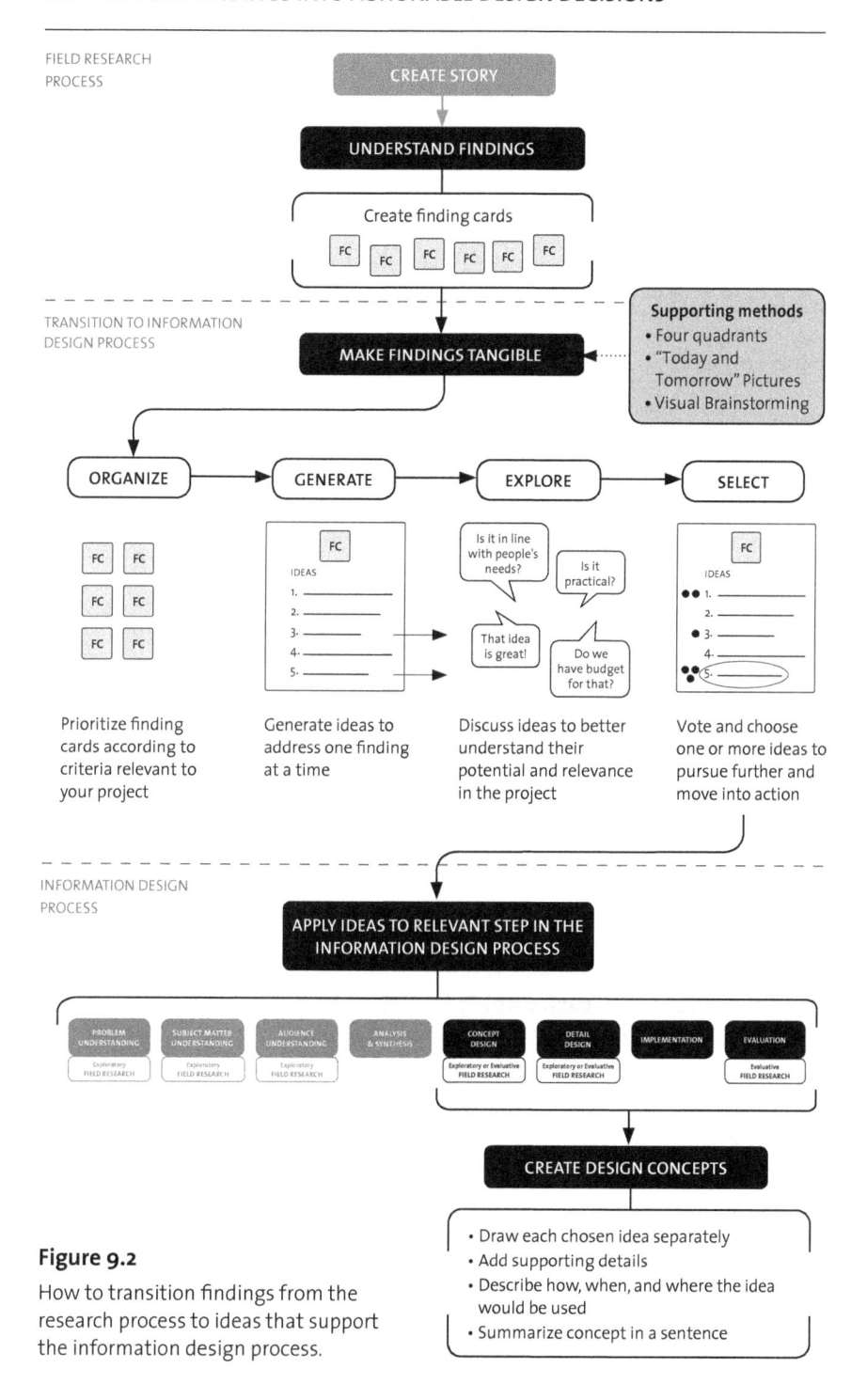

Figure 9.2

How to transition findings from the research process to ideas that support the information design process.

Bridging to design | 185

indicator of something that people are missing, a problem, something that could be improved, or an opportunity to create something new.

Let's analyse the following two findings as an example:

EXPLICIT FINDING	
Finding 1	Intended audience struggles to find their way in the museum using the new wayfinding system and has expressed that they don't understand the icons.
Interpretation	You can infer that there is a problem in the wayfinding system because the intended audience can't find their way. You can also infer that the problem seems to lie with the icons, as the intended audience doesn't understand them. In this case, one possible solution could be to redesign the icons.

LESS EXPLICIT FINDING	
Finding 2	Intended audience doesn't understand how to use the prescribed inhaler because the instructions included in the box were described as "unclear".
Interpretation	You can infer that there is a problem with the instructions. But you can't be sure specifically what is unclear: was it the formatting? Was it the terminology used? Was it the amount of information? Each of these questions triggers many different possible solutions to address the same finding.

One way to start making sense of your findings is creating **finding cards**. Each finding represents something you learned from the study: a point of view, a need, a value, a behaviour, a struggle, or a response to a design idea or prototype. Each finding card is a succinct statement about a single thing you have learned. To create finding cards:

1. **Make a list:** List everything you have learned from the study on a sheet of paper. This helps you collect your thoughts in one place.

2. **Transcribe to card format:** Write each finding in sentence form on a single card (e.g. using index cards or big sticky notes; Figure 9.3). The card format allows you to organize and group findings in different ways.

3. **Get familiar with each card:** Spread all finding cards on a table or wall to have a better understanding of what you found.

Make findings tangible

At this stage, findings are statements that describe participants' experiences. Findings become actionable when ideas are tangible, such as in the form of design concepts. There are many techniques you can use to transform statement findings into more tangible action items; most of them involved the following sequence of tasks that you can do with your team:

Figure 9.3
Finding cards from a contextual interview study.

- **Organize:** Give order and prioritize finding cards according to criteria relevant to your project. This helps determine which finding/s the team should focus and act on first. If the study is small and you only identify two or three findings, you won't have to spend much time organizing them. But if you identify more than ten relevant findings, determining where to start could be hard.

- **Generate:** Focusing on one finding at the time, generate ideas to address it, such as in a brainstorming session[2] or other generative activity.

- **Explore:** After idea generation, discuss some of the ideas to better understand their potential and relevance with the project.

- **Select:** Choose one or more ideas to pursue further and move into action. To make this decision, the team should select ideas based on specific, previously agreed criteria (e.g. relevance, practicality, implementability, and enthusiasm). Each person votes for, for example, their three or five favourite ideas, then discussion continues until everyone agrees how to move forward: here, which idea will be executed first.

2 Kelley, T. (2001) *The Art of Innovation*, Crown Business.

FOUR QUADRANTS TEMPLATE

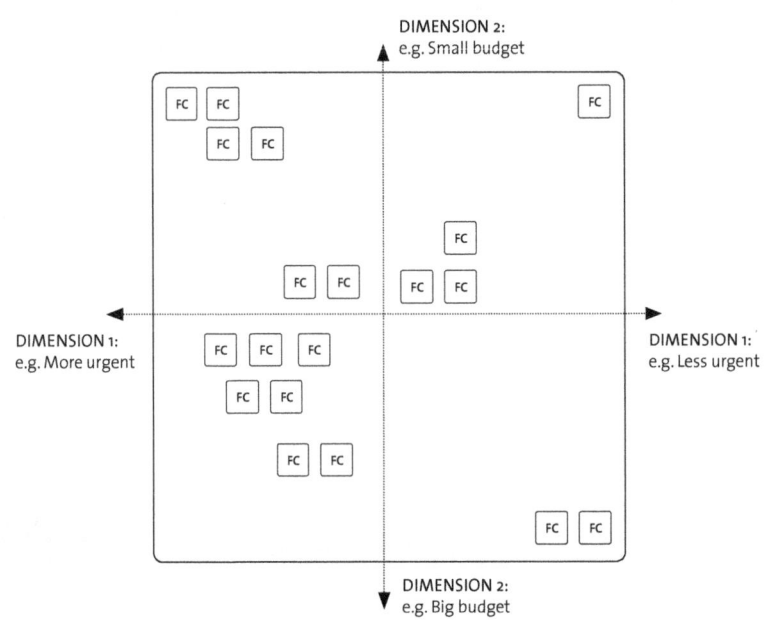

Figure 9.4
Finding cards organized within the four quadrants tool.

The following sections discuss three approaches: **Four quadrants, Today and Tomorrow Pictures** and **Visual Brainstorming.** Some steps are similar across the three, but each offers a unique way to working with field research findings. Depending on the findings you have, the size of your team, and the moment in the design process, one approach may be more useful than the others.

Four Quadrants

This tool can be used to give order to findings using a matrix often called *Four Quadrants*[3] *chart*. The matrix is based on two specific dimensions, such as importance and urgency, or cost and time, used to organize findings. Then you can identify high-priority findings and generate ideas. To use this tool:

- **Organize:** Create a 2×2 matrix with opposing characteristics or dimensions on each end of the spectrum of each axis (e.g. less vs more urgent

[3] This tool is also often used early in the process to synthesize data – IDEO, The global design company: https://www.ideo.com/ [Accessed 10 April 2017].

188 | Chapter 9

findings, and big budget vs small budget; Figure 9.4). You can create this matrix on a wall, whiteboard, or flipchart paper. Plot findings cards in each quadrant, according to where they fall along the spectrum. This sorting step should be done together with your team to ensure all key aspects of the project are considered for each finding.

- **Explore:** Once all finding cards are plotted, analyse and rank them in each quadrant, to determine which should have higher priority.
- **Select:** Narrow down findings to two or three that the team can start working on first and develop design concepts.
- **Generate:** Organize a brainstorming session to find ways to evolve each finding as a design concept.

This tool can be particularly useful if you identified a large quantity of findings, and the team has many constraints to work with (e.g. low budget, tight time frame, specific materials). In contrast to the next two approaches, the strength of this tool is on the organization of findings rather than on the generation of ideas.

"Today and Tomorrow" Pictures

This technique makes findings tangible by creating Today and Tomorrow pictures. The **Today picture** represents the current reality and illustrates a persona's story or situation based on data and findings. The **Tomorrow picture** represents a 'desired future state'[4] and expresses how the same story or situation would change and improve after any identified problems or struggles have been addressed by, for example, a new design or a smoother process (Figure 9.5). Then, the goal is to identify what is needed to move from one picture to the other. Your team can use this technique in a workshop internally, or invite clients and intended audiences to contribute to the session.

- **Organize:** Determine what persona's story or situation you want to focus on and use to create the Today picture. Always based on research learnings, list elements that are important for your persona that can help accurately illustrate that story, such as tasks, events, processes, struggles, or events. Use the *Five Ws + One H questions* to help identify all key dimensions of the story that should be included in the picture.
- **Generate:** Create the Today picture to visualize the chosen story using the list you created in the previous step and identified dimensions. The Tomorrow picture will be based on the same initial story or situation,

[4] Treffinger, D.J., Isaksen, S.G., & Stead-Dorval, K.B. (2005) *Creative Problem Solving: An Introduction*, 4th edn, Waco, TX: Prufrock Press Inc.

"TODAY AND TOMORROW" PICTURES

TODAY PICTURE

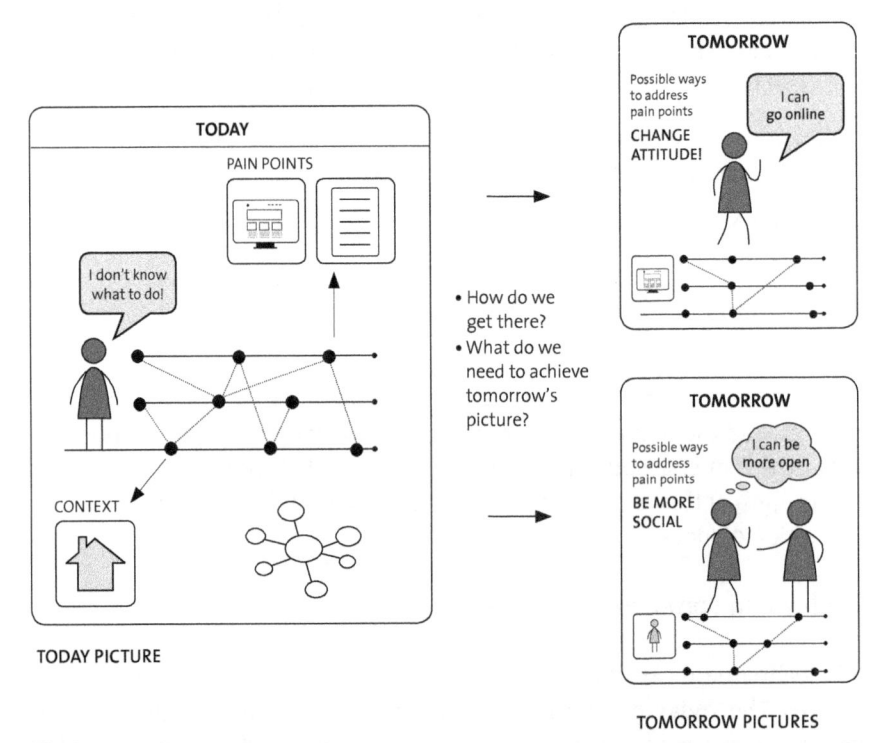

TOMORROW PICTURES

Figure 9.5
At the beginning of the session create a Today picture based on selected findings from the field study that you want to focus on first. Then each team member creates a Tomorrow picture to envision how pain points indicated by findings could be addressed. The team discusses each Tomorrow picture to better understand the changes each person is proposing. The next step is generating ideas to execute some of those changes.

but each team member can contribute with their own vision—that is, ways of improving the current situation—and create their own picture as long as improvements are rooted on actual motivations, desires, or intentions reported by the persona. You will end this part of the session with multiple Tomorrow pictures.

- **Explore:** Pictures are then discussed, evaluated, and, if relevant, combined. The most important criterion to assess a Tomorrow picture is having a clear connection to the intended audience's needs. If your team has created a large amount of relevant ideas, you can create an affinity diagram to organize them into groups, based on similarities. For example, create a list of all aspects of the Tomorrow pictures that work well, another list specifying problems, and a third list describing what you would need to make each new vision happen. Items described in this last list are the team's *action items*.

- **Select:** Decide on which Tomorrow picture/s the team will focus on first, and identify two or three action items for the team to start working on first and develop design concepts.

Action items will be the starting point for the team to generate design concepts. The team can make changes to the selected Tomorrow picture/s, or generate ideas to address identified problems.

Visual brainstorming

Visual thinking and sketches help externalize ideas and connect them to the data[5]. This technique helps transform data into ideas by visualizing concrete ways to address each finding. Ideas can be sketched as storyboards, journeys, or just single drawings. You can sketch by yourself or with your team, or organize a workshop with stakeholders and intended audiences (Figure 9.6).

- **Organize:** Read through your finding cards and identify those that represent something tangible, such as a scenario or accomplishing a task. You and your team will start working with these. Identify what persona and needs each finding card belongs to. Keep the persona handy so you can access its values, common pain points, goals, and motivations.

- **Generate:** Like with the Tomorrow picture, generate ideas to address each selected finding, but sketch one idea at the time in a single page. Add many details and specifications to each sketch to help make each idea more tangible. Working with a scenario in mind helps trigger ideas more in line with the persona's characteristics, and explore more realistic tasks, activities, or situations that they may encounter and how an idea would support them. You can create many sketches representing many different ideas to address each finding, or many sketches to represent different aspects of a same idea, or both.

- **Explore:** Go through the resulting sketches, discuss and decide which ones more clearly communicate the idea, have more level of details, present more relevant ideas, or better respond to any other assessment criteria in line with your project. If you are working with a team, each member can vote their preferred ideas and sketches.

[5] Rojas J. (2017) Etch A Sketch: How to Use Sketching in User Experience Design, *Interaction Design Foundation* [online], Available: https://www.interaction-design.org/literature/article/etch-a-sketch-how-to-use-sketching-in-user-experience-design [Accessed 19 December 2017].

Figure 9.6
Visual brainstorming session. Designers and artists from *Escuela de Arte y Superior de Diseño* (Soria, Spain) sketching ideas to address findings.

- **Select:** Choose two or three ideas for the team to start working on first and develop design concepts.

The visual nature of this technique can make clients or stakeholders feel uncomfortable having to express their ideas with drawings. But if they are willing to participate, their input can be really valuable. One way to help them feel more at ease with drawing is pairing each client with a member of the design team, and work together to sketch ideas.

From ideas to design concepts

After a brainstorming session, a Today-Tomorrow workshop, or a pile of sketches, you will have many ideas. If you are at the end of the middle stage or at the late stage in the information design process, these ideas can develop into **design concepts**. These represent the underlying logic, structure, thinking, and reasoning of a design—often presented as a set of recommendations, features, or design principles—that your team will use to develop design prototypes.

Unlike sketches, design concepts have more granularity and details. To develop ideas into design concepts, as a team draw each idea separately in a large sheet of paper, and add details (as many as you can) about structure, features, materials, components, size, dimensions, colours, and context of use. Also describe how each idea would be used, when and where. Summarize the essence of the design concept in a few words as a tagline. For example, in the Case Study 1: Carnegie Library project, the MAYA team organized three mixed teams including librarians, designers, and architects to work independently with an assigned persona and scenario, and generate one design concept. After a week, each team presented their design concepts: a clearer wayfinding strategy, a more inviting librarians' desk, and better education for users about library processes. Then, the MAYA team worked on testing and developing each design concept into a prototype.

Support the information design process

As previously mentioned, field research findings have different applications, depending on the information design process stage (Figure 9.7). In earlier stages—problem, audience, or subject matter understanding—findings can inspire the design team during idea generation, while also helping build empathy[6]. Further into the process—concept design stage—findings can also provide inspiration but, in this case, to change direction or improve a design concept. Later in the process—design detail and evaluation—findings indicate prototype areas needing improvement.

Let's expand each information design process stage, in which findings can be used, by describing the most common applications:

Early stage: empathy and inspiration

Define direction: Use findings to break down a complex topic into subject areas and determine an initial direction for the project. Emerging themes could be indicators of key or most relevant subject areas within a complex topic. Particularly when dealing with unframed challenges, the beginning of the information design process demands greater use of field research findings to start defining focus and identifying potential audiences.

Identify intended audience's needs: Use findings to create personas and scenarios, and gain a more complete picture of the intended audience's needs, struggles, characteristics, and everyday life experiences. Personas help determine criteria for decision-making throughout the rest of the process. For example, when creating an interface for people within an *Experts Persona* (people that have high technological expertise and a deep understanding

---

6 Sleeswijk Visser, F. (2009) *Bringing the everyday life of people into design*, PhD thesis, Delft University.

CONCEPTUAL DESIGN

```
PROBLEM            SUBJECT MATTER      AUDIENCE           ANALYSIS
UNDERSTANDING      UNDERSTANDING       UNDERSTANDING      & SYNTHESIS

Exploratory        Exploratory         Exploratory
FIELD RESEARCH     FIELD RESEARCH      FIELD RESEARCH
```

We can use findings to define direction!

Or to gain a more complete picture of the audience!

Or to support idea generation!

EARLY STAGE: Empathy and Inspiration

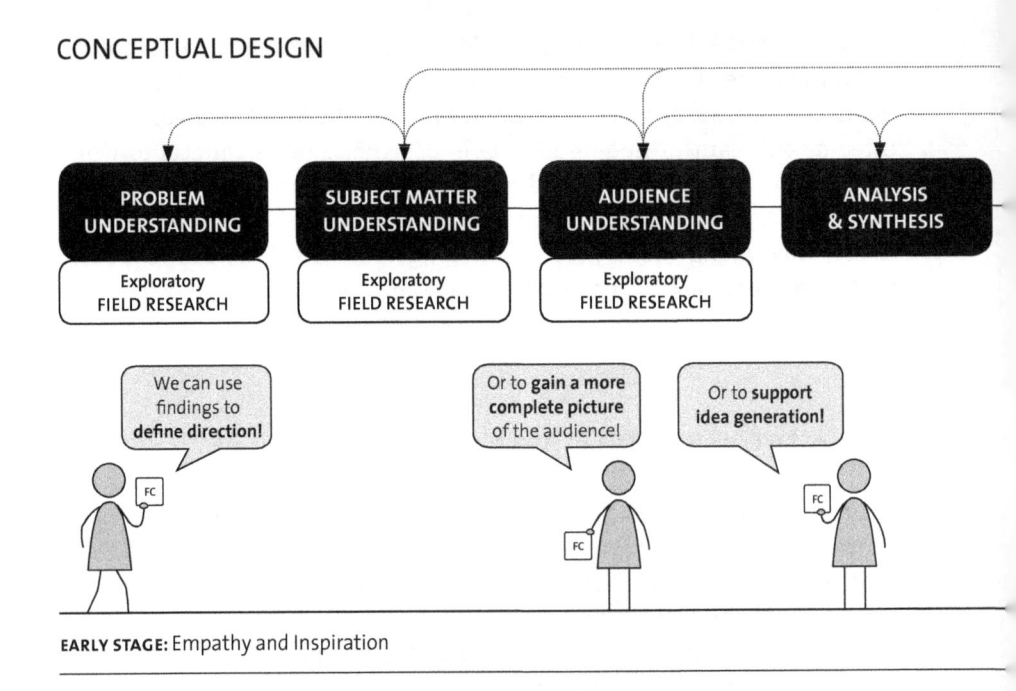

Figure 9.7
Use of findings at different stages of the research-led information design process.

of healthcare), you may tailor language and design features to that persona, through advanced terminology, extended functionality, and minimal tutorial guidance. Emerging themes may indicate how each persona performs tasks or stress those steps in a process that could be supported by a new design. For example, if you create a journey, this provides a picture of concrete tasks that participants perform at specific steps of a given process. Each step may represent opportunities to design ways to support those tasks.

Support idea generation: Use findings as a source of inspiration, to give direction to brainstorming sessions and generate more targeted ideas to develop design concepts. Use identified inconsistencies between what participants said, did, or heard as indicators of latent needs. These could be opportunities to create something new or improve an existing design.

Middle stage: inspiration and improvement

Evolve a concept or idea: Use findings to expand an idea. Findings may suggest ways that you and your team didn't imagine in which an idea could

PROTOTYPE DESIGN

```
CONCEPT          DETAIL          IMPLEMENTATION    EVALUATION
DESIGN           DESIGN

Exploratory or Evaluative   Exploratory or Evaluative              Evaluative
FIELD RESEARCH              FIELD RESEARCH                         FIELD RESEARCH
```

"Later we can use findings to **evolve concepts!**"

"Or to **validate** concepts or ideas"

"And to **optimize** a solution!"

MIDDLE STAGE: Inspiration and Improvement   **LATE STAGE:** Improvement

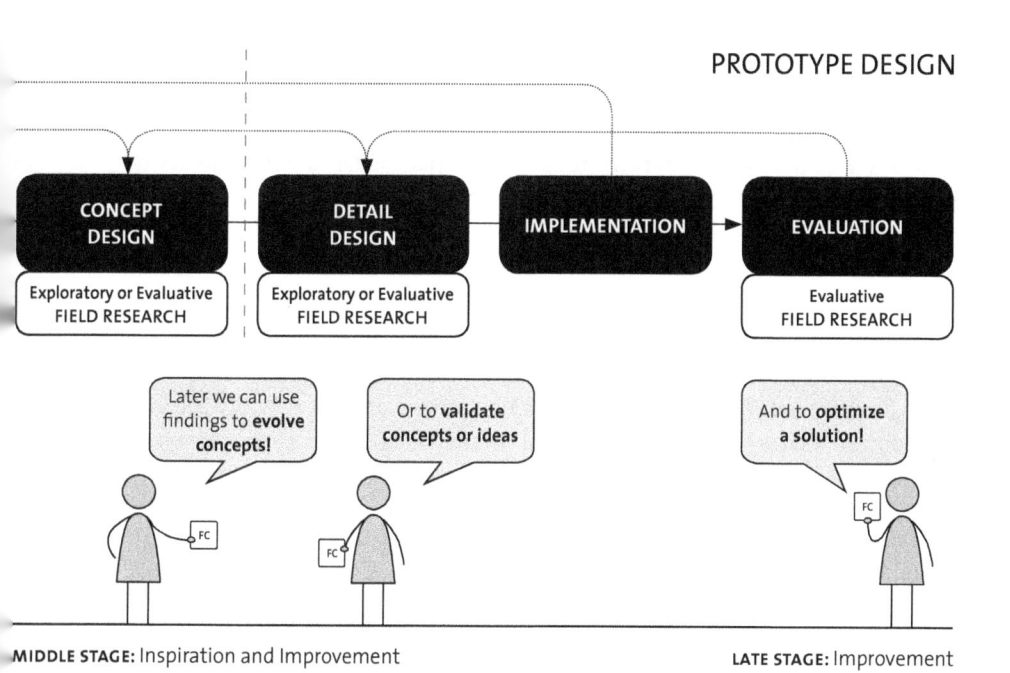

---

grow to better support audience's needs. Findings may also indicate specifics you could add to the idea to improve both its functionality and appealing, such as materials, accessories, uses, format, or dimensions.

Validate a concept or idea: Use findings to refine an idea or discard it and take a different direction. Findings may indicate possible risks or offer unexpected directions that you or your team hadn't initially considered. Findings may also reveal that the intended audience doesn't find an idea engaging or interesting or that, even though they do find it interesting, they won't use it because the idea doesn't align with their needs in that particular moment; for example, it is too complex, hard to use, time-consuming to understand, etc.

Late stage: improvement

Optimize solution: Use findings to understand why a prototype works or doesn't work. It is important to pinpoint what users reported as interesting and why they expressed an interest in using it. These features or dimensions could be replicated later in further iterations of the prototype.

For example, each theme that emerged from the analysis—as the result of creating an affinity diagram—may represent a design issue in a prototype that needs improvement, or perhaps all themes may point to different dimensions of the same design issue. As another example, specific tasks shown in a storyboard performed by participants, while interacting with a new design, may indicate which steps work as planned and which need improvement. Often, *pain points* indicate opportunities for intervention, and *touchpoints* represent ways in which interventions may be delivered.

If you generate a new design concept or put together a list of improvements for an existing project based on field research, but you can't make connections back to the findings—that is, explicitly articulate how data will inform your design decisions, and specifically how those decisions will translate into design features or functions—you aren't using data and learnings to support the information design process.

If you generate design concepts and can clearly articulate their link to data and findings, your next step is executing those concepts moving through the last steps in the information design process: give structure, determine layout, define function, and add visual design.

10 Putting it all together

When limited time and budget are available, you can take the role of information design researcher. Field methods discussed here can be adapted and applied across a range of intensities, varying from single sessions with a handful of participants to larger research projects with many participants. Small, one-day field studies can provide the insights you need and are as useful as four-week long studies. Both can reveal interesting and revealing data about people's lives and behaviours.

With practice, you and your team will discover your own effective and efficient ways to conduct field studies. You may prefer working by yourself or with a team, with some of you working in parallel, meeting daily or twice or three times a week, to share observations, ideas, and preliminary insights, throughout the course of the study or just coming together to discuss the analysis. One of you may be responsible for recruiting participants, another for gathering data, yet another for the analysis of data, or you may decide that the team should do everything together. Either way, being part of the research process and sharing and discussing insights will influence each of you to return to the information design process with a more empathetic understanding of the needs of your intended audience, because you will be immersed more deeply in their context, rather than passively designing based on assumptions.

Myths and assumptions, and understanding the world only through a positivist paradigm, are probably the most common barriers hindering the adoption of field research in information design practice. The more aware you are of them, the more you can objectively determine whether the use of this approach is what you need to successfully complete the project at hand.

If you decide to conduct field research as part of your information design project, take the guidance discussed throughout the previous chapters to design the study, select the methods to gather data or evaluate your designs in the field, and follow the tips to make sense of the data. Figure 10.1 summarizes all key concepts explained, provides estimated time frames, and indicates most common resources needed at each step to help you make your field study work around the constraints of your project.

With practice, you will gain confidence in field research. You have already taken the first steps in that direction.

OVERVIEW OF FIELD RESEARCH IN INFORMATION DESIGN

CHAPTER 4

CHAPTERS 5 & 6

FIELD RESEARCH

We don't know if our **audience needs** an app

Let's do **field research** to find out!

TEAM OF INFORMATION DESIGN RESEARCHERS

INFORMATION DESIGN PROCESS STEP

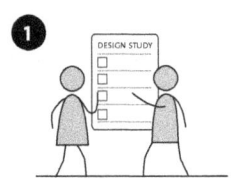

STUDY PARTICIPANTS

DESIGN STUDY
(3–5 days)
- Define goal
- Determine focus
- Recruit participants
- Select data collection methods
- Select data analysis methods
- Plan how to share findings

GATHER DATA
(2–10 days)
- Observations
- Contextual interviews
- Contextual inquiry
- Diary studies
- Design probes
- Collaborative workshops

ESSENTIAL RESOURCES

- **Participant recruitment materials, instruments** for data collection, and **forms.**
- **Participant compensation** depending on the study. It can range from a cup of coffee to gift certificates of $50 or more.

- **Tools and equipment** for data collection (notebooks and digital devices) and the **skill** to use them.

ESTIMATED TIME FRAMES

- **1–2 days** to get familiar with the topic and plan study.
- **1 day** to write information sheets and consent forms.
- **1–2 days** to create and pilot tools, instruments, and materials.

- **Roughly 45–90 minutes per participant,** given that data collection greatly depends on the method used and sample size.

Figure 10.1

To help you plan your field study, follow the steps described here and use the estimated time frames and resources to get a better sense of what each step involves.

HELPFUL TIPS

Allow time for ethical approval, which depends on the study's level of sensitivity and internal policies of your organization.

Manage your daily interviews: more than 4 per day can be exhausting.

CHAPTER 7　　　　　　CHAPTER 8　　　　　　CHAPTER 9

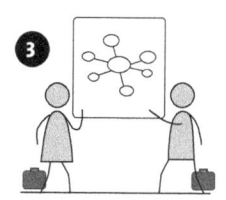

MAKE SENSE OF DATA
(3–5 days)
- Five Ws + One H
- Visual content analysis
- Affinity diagram
- Empathy maps
- Needfinding
- Personas
- Visualizations

SHARE FINDINGS
(3–7 days)
- Visual story
- Personal cardset
- Personas and scenarios
- Storyboards

MAKE FINDINGS TANGIBLE (1–2 days)
- Four quadrants
- "Today and Tomorrow" Pictures
- Visual brainstorming

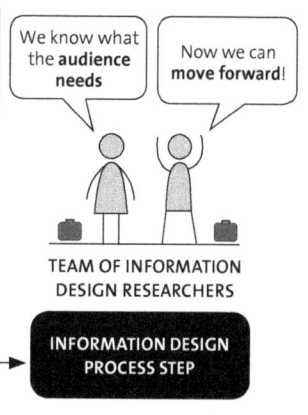

TEAM OF INFORMATION DESIGN RESEARCHERS

INFORMATION DESIGN PROCESS STEP

- **Workspace** to pin up or spread out field data and enable visual connections.
- **Whiteboards** and **sticky notes** to capture thoughts during the analysis.

- **Deliverable format and specifications** based on the audience and information design process step.

- **Workspace** to pin up or spread out finding cards.
- **Writing and drawing materials:** whiteboards, flipchart paper, index cards, markers, sticky notes, etc.

- **1–2 days** to prepare your data for analysis.
- **2–3 days** to analyse data, but it will depend on use of digital tools, people involved in analysis, focus of the analysis, etc.

- **2–5 days** to create a visual story for an external audience.
- **1–2 days** to create personal cardsets, personas, and scenarios or storyboards for internal or external use.

- **1–2 days** to organize findings, and generate, explore, and select ideas.

Be patient and don't skip steps! Trust the process.

Don't be afraid to break conventions if an alternate way is clearer.

Findings aren't solutions! They represent problems, not how to solve them.

Putting it all together | 199

PART IV:
Case studies

11 Field research in information design practice

This chapter discusses five cases from different information design companies and organizations that have used some form of field research as part of their design process. Cases presented here are twofold: they illustrate ways in which field research can support various moments in the information design process and highlight the increasingly ambiguous and unframed cross-disciplinary challenges facing information designers today. Regardless of framing or level of complexity, all have involved a cross-disciplinary approach.

In these cases, the use of field research contributed to the design of successful outcomes, and research methods have been adapted to each project's time frame and set of constraints. Table 11.1 indicates the methods used in each case study to address the respective information design problem types. Some of the case studies exemplify more traditional information design projects (Case Studies 1, 3, and 5). Others represent unframed information design challenges involving collaboration with service design, wayfinding, architecture, and web design professionals (Case Studies 2 and 4). Together, they capture the wide range of activities involved in the information design process. Some of the solutions in the projects discussed here didn't involve the design of a tangible artefact but the redesign of a service experience or a programme to develop more supportive human interactions.

These five cases represent the various ways in which field research can greatly benefit information design practice. Each case opens with an explanation of the overarching goal of the project, the team, and the role of research in the process.

	INFORMATION DESIGN PROBLEM TYPE	METHODS USED IN FIELD RESEARCH
CASE STUDY 1: The Redesign of the Carnegie Library of Pittsburgh	**Experiences, Services & Processes** SOLUTION: Organizational changes, a clearer wayfinding system, and the adoption of new technology *Interactions / Maps & Signage*	• RESEARCH: Non-participant observations, walk a mile in library users' shoes (participant observation), shadowing, contextual interviews • ANALYSIS: Direct-experience storyboards, personas, use case scenarios, and concept maps • DESIGN: Co-creation workshops • REFINEMENT: Pilot formative evaluations
CASE STUDY 2: Legible London	**Systems, Experiences, Services & Processes** SOLUTION: A city-wide wayfinding system *Systems / Maps & Signage*	• Desk research • Field interviews (with public and stakeholders) • Mental mapping • Observational studies • Quantitative studies • Functionality tests • Formative overt and covert evaluations • Before-after implementation evaluations • Key informants • Visual content analysis
CASE STUDY 3: Vendor Power Guide	**Artefacts** SOLUTION: A print educational resource *Visual explanations*	• Desk research • Non-participant and participant observations • Contextual interviews • Formative overt evaluations
CASE STUDY 4: A Better A&E	**Systems, Experiences, Services, Processes & Artefacts** SOLUTION: On-site information and communication system, training programmes, a publication, and website *Systems / Maps & Signage / Documents*	• Desk research • Exploratory ethnographic research (observations and interviews) • Formative and summative evaluations • Customer journey • Workshops • Non-participant observations • Incidents analysis
CASE STUDY 5: To Park or Not to Park	**Experiences, Services & Processes** SOLUTION: A template to design a user-centred parking sign *Maps & Signage*	• Desk research • Non-participant observations • Surveys • Contextual interviews • Formative and summative evaluations • Participative design

Table 11.1 Overview of case studies discussed in the chapter, the information design problem type, and methods used in the field research studies.

CASE STUDY 1:
The Redesign of the Carnegie Library of Pittsburgh

Pittsburgh, PA, US, 2002–2004

Project team

The team for this project[1] was composed of librarians and library directors responsible for setting the vision, employing vendors, and finding funding; architects responsible for renovating the library; and designers[2] from MAYA[3], a design consultancy and innovation lab, responsible for understanding library users and visualizing ideas.

Project goal

The goal was to 'change the public's perception of [The Carnegie Library of Pittsburgh (USA)] as a dark, forbidding place full of old, irrelevant books to one of a bright, inviting place, teeming with up-to-date, relevant information'[4].

Research approach and methods

- Observational studies: Non-participant observations, walk a mile in library users' shoes (participant observations), and shadowing
- Contextual interviews
- Direct-experience storyboards
- Personas, use case scenarios, and concept maps
- Co-creation workshops
- Pilot formative evaluations

---

1 Based on Pontis, S. & Babwahsingh, M. (2016) Improving information design practice: A closer look at conceptual design methods, *Information Design Journal*, 22(3), 249–265.
2 McQuaid, H.L., Goel, A., & McManus, M. (2003, June) When you can't talk to customers: Using storyboards and narratives to elicit empathy for users, in *Proceedings of the 2003 International Conference on Designing Pleasurable Products and Interfaces*, ACM, 120–125.
3 MAYA, a company of The Boston Consulting Group, Inc. (Pittsburgh, Pennsylvania, USA): http://maya.com/
4 McQuaid, H.L., Goel, A., & McManus, M. (2003, September) Designing for a pervasive information environment: The importance of information architecture, in *Proceedings of the British HCI Conference*, Bath, UK.

Figure 11.1

Research outputs. Notes from brainstorming sessions and interviews with librarians.

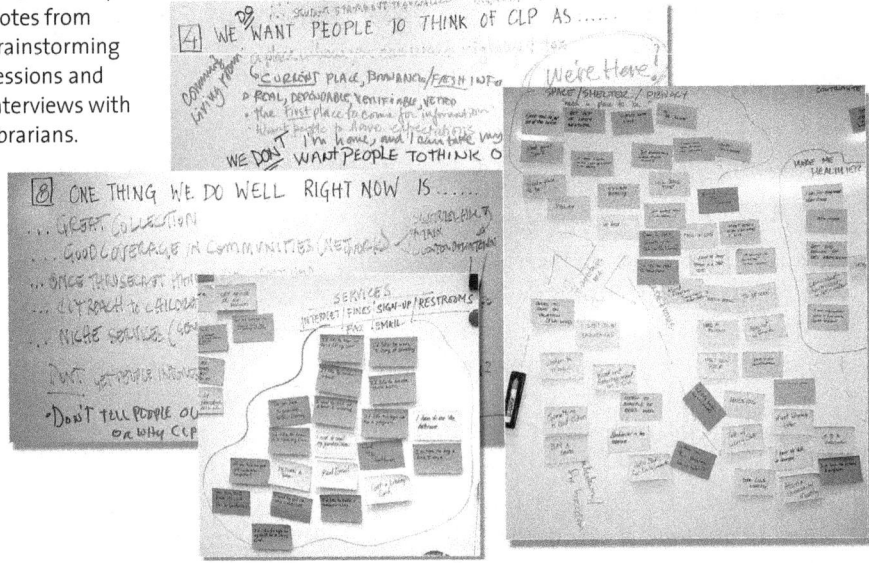

Process

MAYA designers divided the project into four stages: research, analysis, design, and refinement and implementation.

Research: To gain a better sense of the problem, designers first had to understand the scope and complexity of the project and the library's organizational structure, and determine who would be interacting with the information and the most frequent kinds of information they were interacting with[5]. MAYA designers facilitated several input sessions with stakeholders[6] and interviewed, observed, and shadowed librarians to identify their key tasks and activities (Figure 11.1). One key constraint of the project was that MAYA designers couldn't talk directly with library users due to privacy concerns. Consequently, the designers decided to use non-obtrusive field methods for half a day to gain an understanding of library users' cognitive and emotional needs; the method used was a type of participant observation called "walk a mile in library users' shoes" (Figure 11.2). Members of the

---

5 McQuaid, H.L., Goel, A., & McManus, M. (2003, September) Designing for a pervasive information environment: The importance of information architecture, in *Proceedings of the British HCI Conference*, Bath, UK; Bell, S.J. & Shank J.D. (2007) *Academic Librarianship by Design: A Blended Librarian's Guide to the Tools and Techniques*, Chicago: American Library Association.
6 McQuaid, H.L., Goel, A., & McManus, M. (2003, June) When you can't talk to customers: Using storyboards and narratives to elicit empathy for users, in *Proceedings of the 2003 International Conference on Designing Pleasurable Products and Interfaces*, ACM, 120–125.

WALKED A MILE IN ONLINE SPACE

Figure 11.2

Research outputs. As part of the research phase, MAYA designers walked a mile in online space: they analysed the library's website as if they were a library user. They annotated screenshots indicating confusing parts and areas for improvement.

WALKED A MILE IN PHYSICAL SPACE

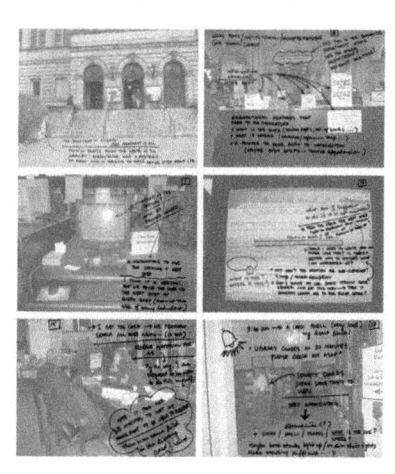

MAYA designers also walked a mile in physical space. Annotated images created from learnings obtained after walking a mile in library users' shoes. Confusion, frustration, and uncertainty emerged as recurrent feelings mostly due to lack of clarity on what to do or where to go, and simple tasks taking too long.

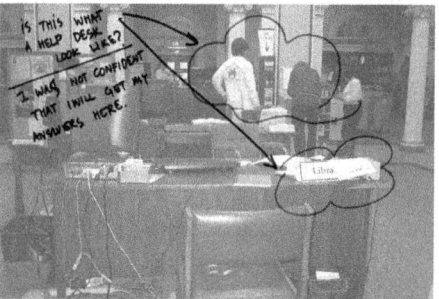

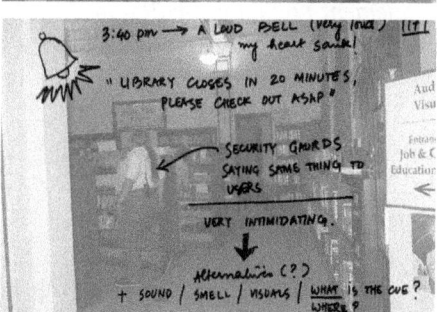

Field research in information design practice | 207

team acted as participant observers, using the library to experience first hand the common tasks that library users undertook daily. As a result of this experience, they compiled and shared their learnings in direct-experience storyboards. Combined, these four methods helped the team create a more complete picture of library users' behaviours, demographic, and flow in accessing and navigating the library. Furthermore, this data allowed the team to better frame the challenge and identify key areas that needed further exploration, such as information-seeking strategies and types of information sources.

Analysis: This stage involved the definition of personas and the creation of use case scenarios and concept maps visualizing the components of the library experience (Figure 11.3). These concept maps had two purposes: to act as a tool to visualize learnings within the team and to communicate findings to stakeholders, such as how library users were interacting with and accessing information. This was useful for identifying the exact moments of the journey where the system was failing to support its users. Key pain points the team identified were unclear system usability, hard-to-find sources, and hard-to-understand information (Figure 11.4).

Design: In order to crystallize findings into design concepts, MAYA used a co-creation approach, employing three "tiger teams": designers, architects, librarians, and other stakeholders were divided into three mixed groups to work for a week on developing design concepts, to address the needs of an assigned persona and scenario. Three design concepts emerged: a clearer wayfinding strategy to help reduce uncertainty and confusion when navigating the library, a more inviting librarians' desk, and better education for users about library processes (Figure 11.5). These concepts helped demonstrate 'how negative experiences could be eliminated, how positive ones could be retained and enhanced, and how new, pleasurable experiences could be created'[7]. The three design concepts were compiled into a set of recommendations to develop an information system for the library.

Refinement and implementation: After this stage, the three design concepts were further explored and piloted. The initial design recommendations were evaluated by library users and then implemented in the library.

---

7 McQuaid, H.L., Goel, A., & McManus, M. (2003, June) When you can't talk to customers: Using storyboards and narratives to elicit empathy for users, in *Proceedings of the 2003 International Conference on Designing Pleasurable Products and Interfaces*, ACM, 120–125.

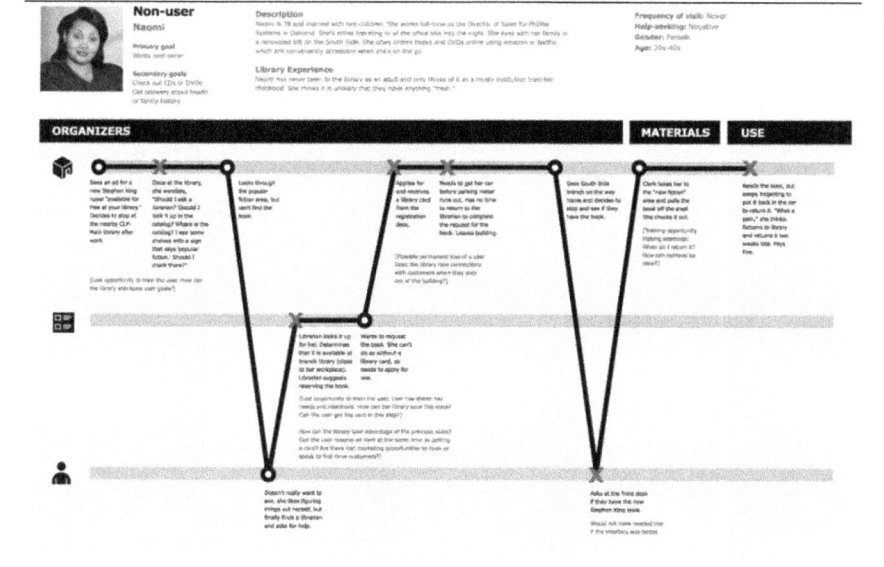

Figure 11.3
Analysis output. Personas and customer journeys representing library users' experiences.

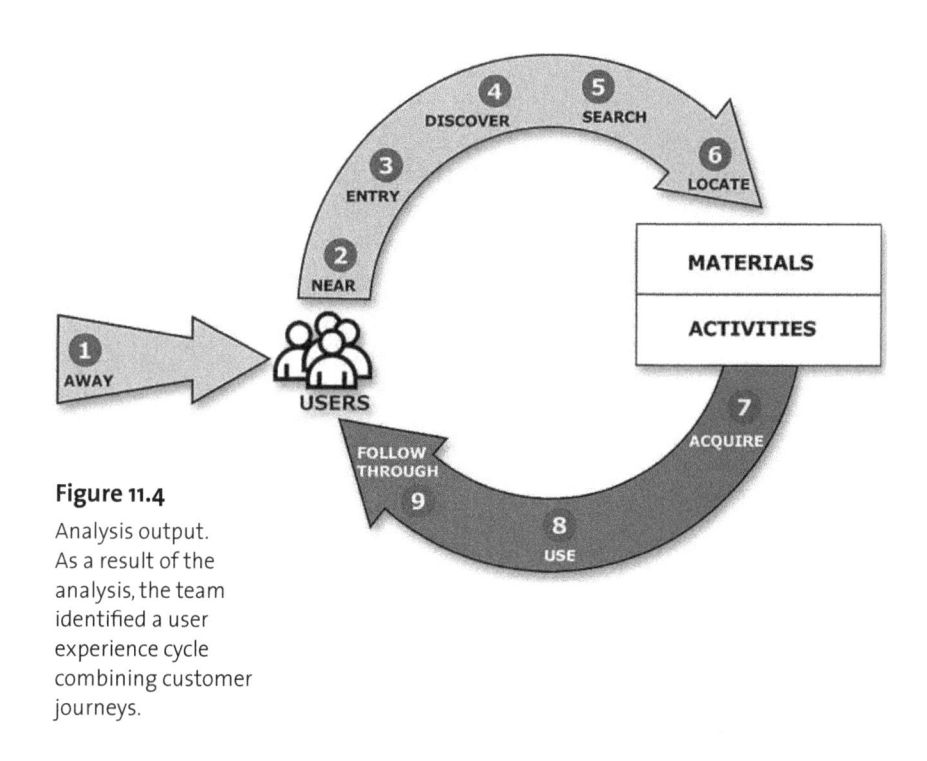

Figure 11.4
Analysis output. As a result of the analysis, the team identified a user experience cycle combining customer journeys.

Field research in information design practice | 209

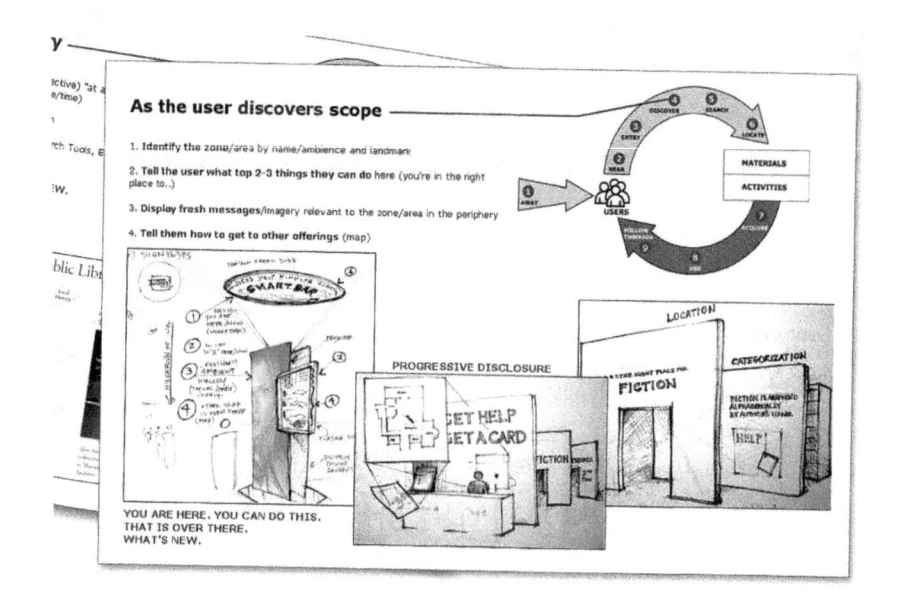

Figure 11.5
Design concepts. Each tiger team created a design concept to address needs of an assigned persona and scenario, and the complete user experience cycle. Case study images from MAYA, a company of The Boston Consulting Group, Inc.

Solution

MAYA Design proposed deep organizational changes, a clearer wayfinding system, and the adoption of new technology[8]. Following completion of this project, the Carnegie Library has become an 'inspiring centre of information and discovery'[9].

---

8 Bell, S.J. & Shank, J.D. (2007) *Academic Librarianship by Design: A Blended Librarian's Guide to the Tools and Techniques*, Chicago: American Library Association.
9 ibid.

CASE STUDY 2:
Legible London

London, UK, 2004–2015

Project team

This long-term project involved the collaboration of several individuals and organizations: the Mayor of London, Transport for London (TfL), the London Development Agency, many London local authorities, research agencies, and Applied Wayfinding[10], an information design consultancy specialized in wayfinding design projects. The team for this project was composed of city planners, designers, and architects, but the core consisted of six in-house designers and urban planners, and one freelance researcher.

Project goal

The goal was to make London a highly walkable city by making it easier for people to move around.

Research approach and methods

- Mix of research methods:
- Desk research (examples of international and national best practices and related projects from various London boroughs)[11]
- Field interviews with pedestrians and stakeholders
- Observational studies
- Mental mapping[12]
- Visual content analysis
- Key informants (in the form of a disability and inclusion review expert panel)

---

10 Applied Wayfinding (London, UK). Available at: http://appliedwayfinding.com/ [Accessed 10 November 2017].
11 Reviewed projects: Bristol Legible City and the UK Road Sign System (1964), Research Business International (2002), MORI study for the London Borough of Islington (2005) as indicated in *Yellow Book: A prototype wayfinding system for London*, 2007, developed by Applied Information Group.
12 People create mental maps of places, routes, and areas (e.g. cities, towns, neighbourhoods) based on points, objects, or landmarks from the physical environment that they remember. Mental maps aren't strictly geographic, but visualize the connections between memorable points, their locations, and routes used to get to them, relevant to people's needs.

- Quantitative studies

- Functionality tests, (overt and covert) formative evaluations, before-after implementation evaluations

Process

In 2004, the Mayor of London, Ken Livingstone, wanted 'to transform [London] into one of the most walking-friendly cities in the world by 2015'[13]. In response to this challenge, designers from Applied Wayfinding[14] suggested an initial assumption: *London is more walkable than you think*. To learn more about this assumption and investigate ways in which walking could be improved in Central London, the Applied team embarked on a long-term project which spanned roughly ten years. At the core of this project, and to guide the design process and inform design decisions, Applied conducted and commissioned a series of field research studies. The following are six key stages where different forms of these studies were conducted, to better understand the challenge and develop and test a solution:

Exploratory field studies (2005): To validate the initial hunch, the Cities team of Enterprise LSE (consultancy of the London School of Economics) was commissioned by the Applied team to conduct an exploratory field research study, to learn how people went from A to B and their knowledge on distances (e.g. how far would they walk?). The study involved 250 interviews with pedestrians, as they left the Leicester Square tube station, and the creation of 150 different mental maps, reflecting people's understanding of the city. As a result, the Applied team created a design concept and an initial prototype for a unified wayfinding system called *Legible London*. Findings were used to show its need and to demonstrate why this solution could help pedestrians find their way in the city and should be developed further.

The next step was to find the budget. Further research into organizational structures (e.g. commercial organizations, council officials from boroughs, federal government representatives, TfL) was used to identify who would be interested in supporting the project.

Observational studies (2005–2006): A series of studies was conducted to strengthen the case. Findings indicated at least 32 different wayfinding systems for pedestrians in London's central Congestion Charging Zone alone. Differences were seen in design (e.g. lack of common style, colours, fonts),

---

13 Legible London. *Yellow Book: A prototype wayfinding system for London*, 2007, developed by Applied Information Group, p. 9.
14 Credentials. Case Study: Legible London. Report, courtesy of Applied Wayfinding.

Figure 11.6
Field research data. This data was gathered from a study to calculate tube and walk times between 49 tube stations in London. The study team calculated travel times using a mix of field research, which involved walking and timing entry and exit times at all 49 stations, and transport travel data provided by Transport for London.

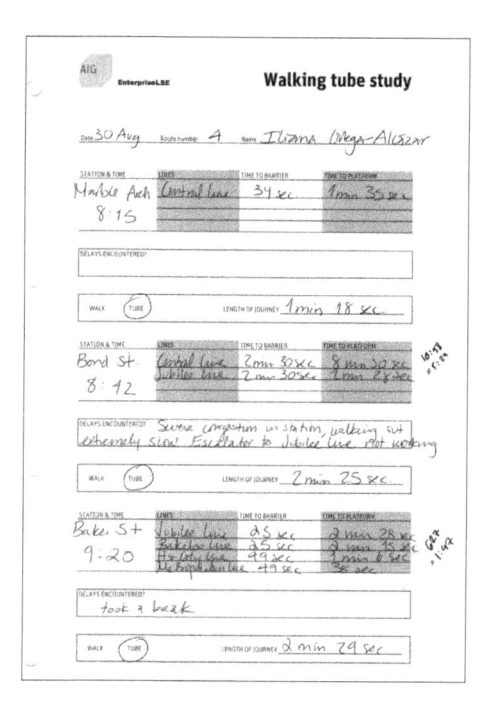

information (e.g. different names, locations), and quality (e.g. different directions, distances). While one study[15] indicated that many commuters used the tube for short journeys (Figure 11.6), a follow-up study helped identify distances in Central London where it was quicker to walk than to take the tube.

Other studies investigated the tangible benefits that a unified, London-wide pedestrian system could generate. The Applied team conducted seven observational studies, asking participants similar questions to those in the first exploratory study, to gauge their thoughts about the challenge (i.e. why people were not walking as much as they could). Responses were video recorded and shown to stakeholders, to support the argument and need for a new wayfinding system. These videos were very well received by the people in the room[16], resulting in TfL joining the project.

Wayfinding study (2006)[17]: An exhibition at the New London Architecture Centre was launched, together with another mental maps study (Figure 11.7). The public were invited to draw their understanding of the city and

15 Walking the Tube. A study into 'walkable' tube journeys in Central London. Conducted by AIG for Transport for London, November 2006.
16 Tim Fendley and Ben Acornley, partners and creative directors of Applied Wayfinding. Phone interview – 16 August 2017.
17 Legible London. *Yellow Book: A prototype wayfinding system for London*, 2007, developed by Applied Information Group.

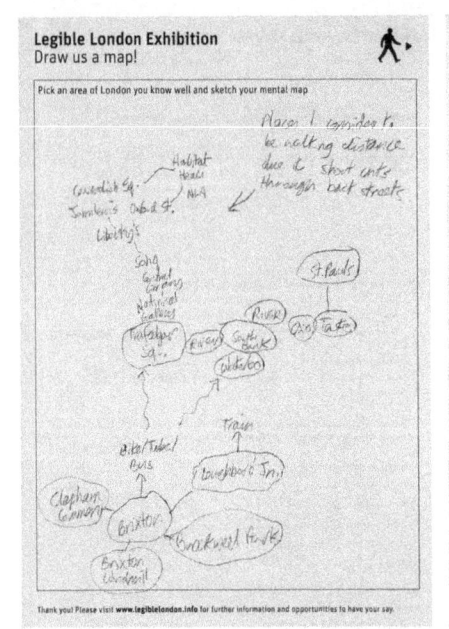

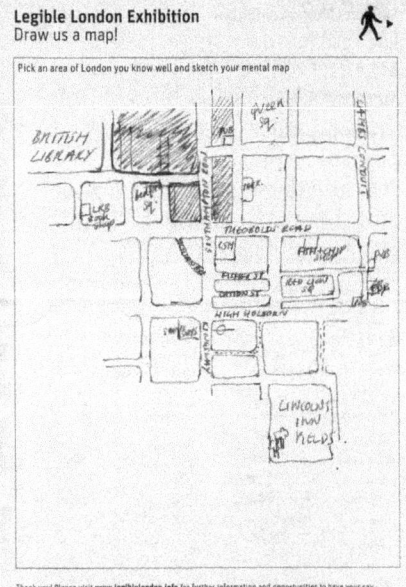

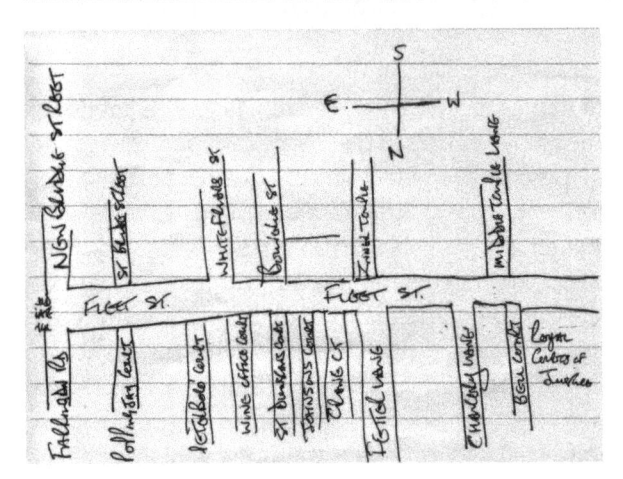

Figure 11.7

Field research data. Mental models created by the public at the Legible London exhibition at New London Architecture.

complete feedback forms on the system concept. This study shed light on which wayfinding cues people would find useful and helped encourage people to get involved and feel ownership of the system. At the same time, an official website about the project was launched (www.legiblelondon.info), inviting feedback through an online poll. Insights helped to identify the needs of local and national government and transport authorities and other stakeholders, and to better understand the needs of the different types of walkers in the city (residents and tourists).

Figure 11.8
Research session. Naming consultation study to determine the most useful language and terminology that the wayfinding system should use.

Following these studies and after funding was secured, 'The whole design of the system only took four months' and involved 'an iterative process: going out every Friday to test the prototype and gather people's inputs and responses'[18]. Throughout these months, the Applied team conducted 'little field studies' in the streets, where some team members went 'under-cover' to Oxford Street and talked to pedestrians. They asked pedestrians for directions, to learn the words and terminologies they used. These findings (e.g. people's words, terms, experiences) provided evidence for making confident content decisions that, combined with the Applied team's design instincts, helped create a robust human-centred system. For example, rather than using official terms or names, which might be language that didn't resonate with pedestrians and that they might feel they were being forced into, the Applied team built the system using words that pedestrians in the study used, such as village names and landmarks, so that they could relate to them. In addition, any names proposed by local stakeholders were first tested and checked with pedestrians before being included in the system (Figure 11.8). As an example, in early discussions about the system, some

18 Tim Fendley and Ben Acornley, partners and creative directors of Applied Wayfinding. Phone interview – 16 August 2017.

Figure 11.9
Design discussions. The Applied team discussing different design options of the Legible London system in preparation for a testing and refining study in the West End.

parties wanted to use *metrics* to show distances between places. However, the Applied team argued that showing distances in *minutes* would be more in line with people's preferences, as 80% of participants described distances from one place to another in minutes. As a formative evaluation, the Applied team mocked up a sign with bespoke pedestrian mapping on it and erected it near their studio. In the space of an hour, around 40 people approached, unprompted, to check it out and provide feedback[19]. To make more informed design decisions, these user studies were complemented with a study focusing on how the brain works, how it cuts corners, and how it recreates spaces in the mind.

As the system designer, Applied drafted and distributed over 30 documents, providing design and implementation guidance, to ensure consistency across pilot studies. In addition, a series of workshops and meetings was organized, to ensure that guidance was understood as planned.

Feasibility studies (2006–2007)[20]: Evaluation studies were conducted in partnership with the South Bank Employers' Group and the London Borough of Richmond, to test feasibility and people's understanding and use of the new system. The choice of two very different locations was intentional, to help test which elements of the system worked universally and to assess where a more tailored approach was necessary. Key considerations tested were 'local distinctiveness, support, costs, management, ownership and maintenance of the system, user requirements, complexity of information planning and benefits to local residents, commuters and visitors'[21].

Testing prototypes (2008): Prototypes of the system were tested in various London boroughs, starting in the West End around the Bond Street tube station, with the installation of 19 on-street pedestrian signs, and then in Covent Garden, Holborn, and Bloomsbury (Figure 11.9). Conceived by the Applied team, the goal of these studies was to assess the effectiveness of specific components of the system such as the placing of signs on streets, place naming use, and sign placement elements.

Summative evaluations and pilots (2008–2009): TfL conducted a series of qualitative and quantitative evaluations of the system, to assess whether it was meeting its objectives—to help people move around the city more easily and with more confidence—and to identify areas for improvement. These evaluations included pedestrian tracking exercises, controlled "mystery shopper"'exercises, knowledge of local areas, and interviews. In addition,

---

19 Phone interview – 16 August 2017.
20 Credentials. Case Study: Legible London. Report, courtesy of Applied Wayfinding.
21 ibid.

Figure 11.10
Final prototype of the Legible London wayfinding system.
Case study images from Applied Wayfinding.

before-after system implementation evaluations were used to measure the system's performance. The main findings showed an increase in the signs' legibility and shorter journeys in the Bond Street area, a decrease in people feeling lost, and high satisfaction levels for the new system.

Solution

The final solution was a walking identity for London in the form of a city-wide wayfinding system, including 500 on-street pedestrian signs, printed tailored walking maps for pedestrians, downloadable and digital maps, smartphone apps, and integrated public transport information (Figure 11.10). Legible London proposes a way to help pedestrians develop stronger mental maps and consequently be more confident about walking to desired destinations and exploring new areas on foot. The new wayfinding system gives people the prompts and building blocks to encourage and support the natural process of mental mapping.

CASE STUDY 3:
Vendor Power Guide

New York, NY, US, 2009

Project team

The Centre for Urban Pedagogy (CUP)[22] coordinated this year-long project as part of the Making Policy Public programme. The team was composed of designer Candy Chang, community advocate (lawyer and former vendor) Sean Basinski, and two project managers from CUP.

Project goal

The goal was to make New York City's (NYC) street-vending rules and regulations clear, understandable, and accessible to street vendors in the form of a print resource.

Research approach and methods

- Desk research
- Non-participant and participant observations
- Contextual interviews
- Formative evaluations

Process

The street vendor population in NYC is highly diverse, originating from Bangladesh, China, Senegal, and Afghanistan. Understanding street-vending rules and regulations has been an ongoing challenge, because English isn't everyone's first language and the rules are so confusing and cumbersome that even police officers sometimes seem to get confused. The team's first step was to learn about the challenges faced by vendors and to gain a better understanding of the rules.

> 22 CUP is a non-profit organization based in Brooklyn, NY, focused on using design to increase the quality of public participation in urban planning to improve urban life in New York City. The organization works with interdisciplinary teams composed of designers, educators, community advocates, and community residents. Each project involves collaborative design work between CUP member partners, designers, and people from the community to develop tools to communicate complex public policy issues in clear and understandable ways. Available at: http://welcometocup.org/ [Accessed 10 November 2017].

To that end, designer Candy Chang and CUP representatives attended monthly street vendor meetings and recorded the discussions. Meetings aimed at informing vendors about common challenges and discussing what they could do to make changes. In addition, the team went to the streets and talked to vendors to learn first hand about their experiences and struggles.

Then the team analysed all collected data and learnings, which helped identify what could constitute relevant content to be communicated in an educational resource. A key step was deciding what type of information vendors needed to know; for this, the team worked on answering the following questions:

> How much would be directed towards street vendors as a much-needed resource, and how much would be an educational/advocacy tool about street vendors and regulation reform? How much would be about clarifying the convoluted regulations into clear graphics and how much would be about showing just how convoluted it currently is?[23]

Once the scope of content to include in the resource was agreed, the next step was to define the layout and determine how information was going to be presented. The designer created a skeleton of the resource, indicating titles and image placeholders which were used to discuss content flow, hierarchies, and general layout. It was important to determine the appropriate order to introduce each rule and regulation, to ensure the message was clear and easy to understand. An important challenge was how to make the information as accessible as possible, considering the wide range of languages spoken among vendors. The team decided, as much as possible, to communicate the rules with visuals and complement them with further explanation, written in five languages: English, Bengali, Chinese, Arabic, and Spanish (Figure 11.11). The resource also included stories about vendors, historical background information, fun facts, and advice for vendors who had problems with the police. For example, one section included this tip: 'Take a photo or a video of your spot. You can use these in court'[24].

Another challenge of the project was to both show the correct content in a visual and understandable way, and communicate "authority". This way, if necessary, vendors could use the resource to discuss any related issues with police officers. After various rounds of experimenting with different visual languages, from photographs to pure illustration, the designer created a

---

23 Chang, C. (2009) Making policy public: Vendor power! *Urban Omnibus* [online], Available at: http://urbanomnibus.net/2009/05/making-policy-public-vendor-power/ [Accessed 10 November 2017] and Knafo, S. (2009) Visual aids for the pushcart world, *The New York Times* [online], Available at: http://www.nytimes.com/2009/04/05/nyregion/thecity/05guid.html?_r=1&scp=5&sq=street%20vendor&st=cse [Accessed 13 January 2018].
24 ibid.

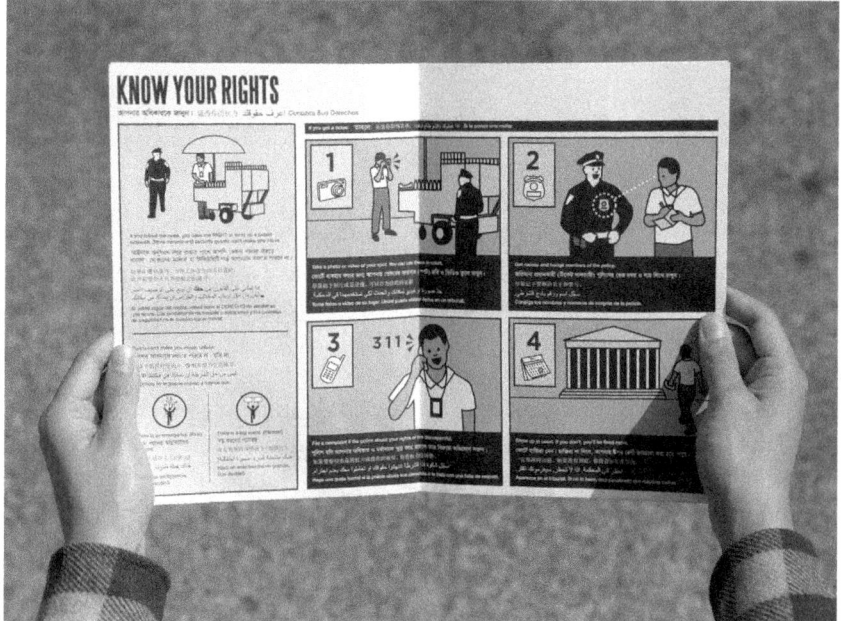

Figure 11.11
Vendors with the final resource. The project team and a group of volunteers distributed the final posters to street vendors in four of the five boroughs in April of 2009.

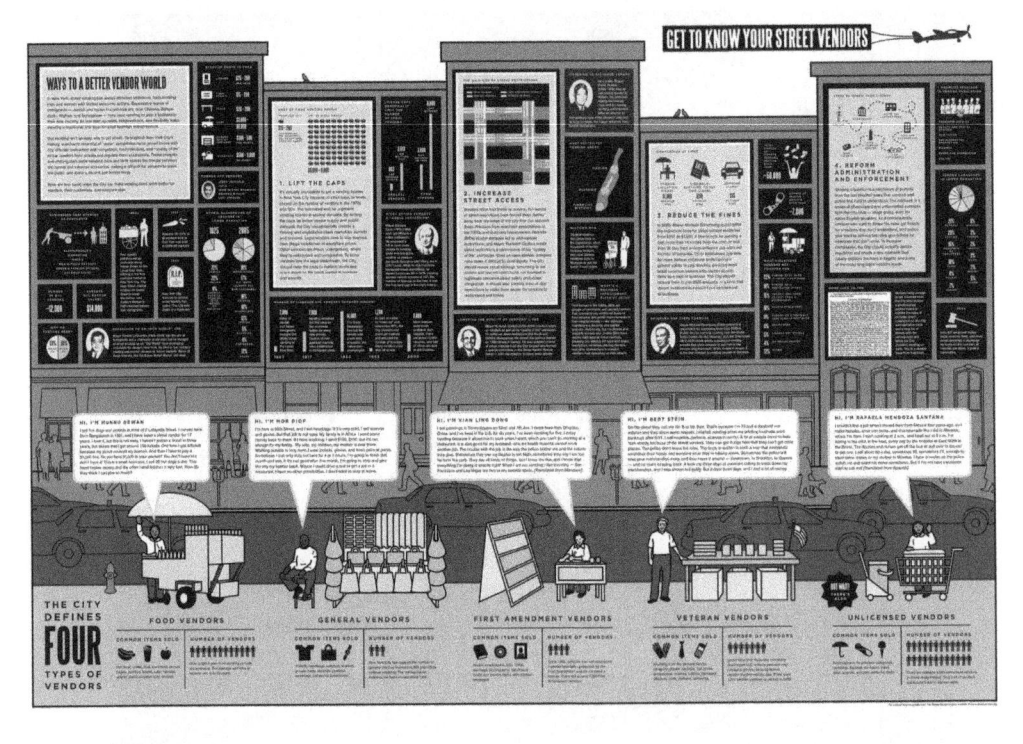

Figure 11.12
Inside the resource. The unfolded resource is a poster, 82cm × 56cm (32" × 22").

prototype using a Chris Ware-inspired style[25] (Figures 11.12 and 11.13) and only using a reduced set of colours: yellow, black, white, and grey. Red and green colours are only used to indicate specific rules allowed by the law.

Testing sessions: The designer was in charge of running the testing sessions and creating questionnaires. The team identified unclear parts in the story and gathered feedback about the illustrated draft resource from 30 vendors in biweekly meetings. These events were already planned for the advocates for other community-related issues and were an opportunity to talk to some attendees about the design, while formally recording the responses. Particularly, vendors' feedback focused on 'clarity, content, symbols, language and text translations'. The team focused on challenging specific assumptions that could hinder understanding such as 'is ">" a universal symbol for "greater

---

25 Chris Ware is an American cartoonist who has published his work in several newspapers (e.g. *The New Yorker, The New York Times, The Village Voice, The Yale Review, Esquire*) and graphic novels [online], Available at: http://fantagraphics.com/flog/artist-bio-chris-ware/ [Accessed 10 November 2017].

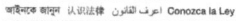

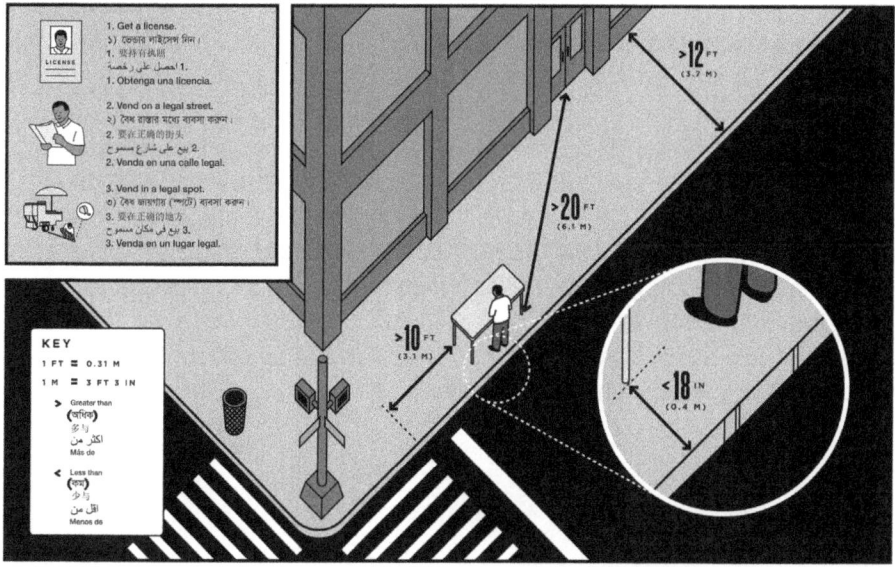

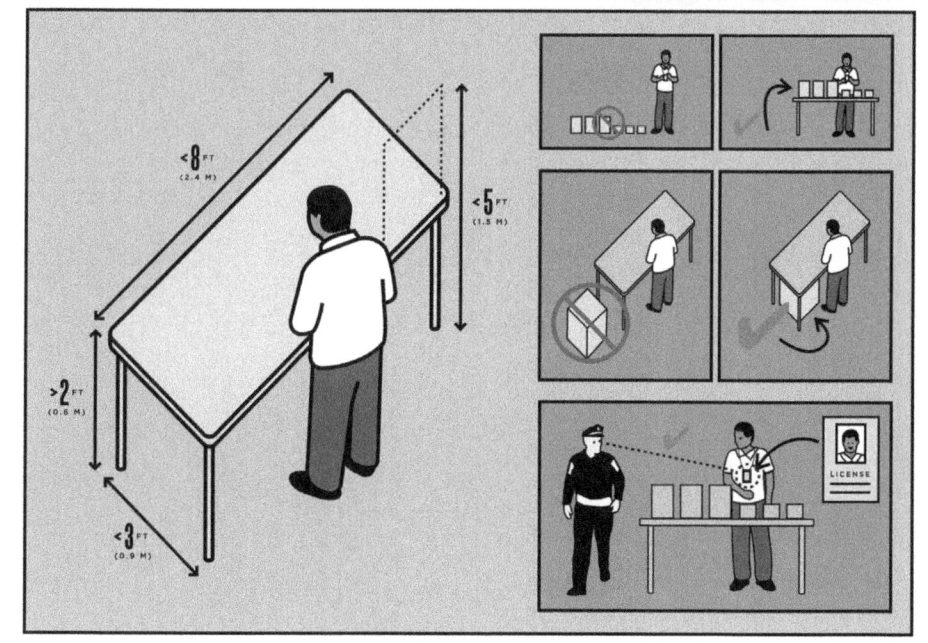

Figure 11.13

Inside the resource. This illustration explains how to get a license, vend on a legal street, and vend on a legal spot. It also provides guidance on how to correctly set up a cart on the street. Case study images from *Making Policy Public: Vendor Power!* © the Center for Urban Pedagogy (CUP), 2017.

Field research in information design practice | 223

than"? Is a green check symbol the opposite of a red x'd circle? Are the abbreviations for feet, inches, and meters clear?'

Solution and implementation

The final 20.3cm × 28cm (8" × 11") educational resource was a fold-out 82cm × 56cm (32" × 22") poster that provided visual explanations of city rules and regulations relevant to street vendors. To publicly launch the poster, CUP organized an event in NYC, where 20 volunteers gave free copies to vendors across the city. To ensure good distribution of the posters, the team spoke with a vendor to learn where in the city were the densest vendor areas and to determine the number of posters needed. In total, posters were distributed to 1,000 vendors across 20 neighbourhoods, including Jackson Heights, Fulton Mall, Grand Concourse, and Harlem.

CASE STUDY 4:
A Better A&E

London, UK, 2011–2013

Project team

This project[26] was commissioned by the Design Council UK[27] and led by design studio PearsonLloyd[28]. The team was composed of designers and strategic thinkers, social scientists, and healthcare professionals from various National Health Services (NHS) hospitals, three NHS Trusts across the UK, an organizational consultant with experience as a former clinical director, and an organizational psychologist.

Project goal

The goal was to create ways to help reduce violence and aggression towards NHS staff in Accident & Emergency (A&E) departments across the UK.

Research approach and methods

- Desk research
- Exploratory ethnographic research (observations and interviews)
- Formative and summative evaluations
- Customer journey
- Workshops
- Watching the service in action (non-participant observations)
- Incidents analysis

Process

To gain a better understanding of the problem, through in-depth desk research, the Design Council first examined existing cases of recent violent and

---

26 A better A&E [online], Available at: http://www.abetteraande.com, and Steven, R. (2013) A better A&E [online], Available at: https://www.creativereview.co.uk/a-better-ae/ [Accessed 10 November 2017].
27 Available at: https://www.designcouncil.org.uk/ [Accessed 10 November 2017].
28 PearsonLloyd Design Studio (London, UK). Available at: http://pearsonlloyd.com/ [Accessed 10 November 2017].

aggressive incidents in UK A&E departments and gained familiarity with previous attempts to tackle similar issues in the health service and in other related services. In addition, the Design Council commissioned studies, involving more than 300 hours of ethnographic research in the partnered NHS Trusts' A&E departments, to identify where and why aggression in hospitals occurred. Findings indicated six personas, representing specific perpetrator characteristics, and nine sets of violence and aggression triggers[29]. Learnings from these research studies were compiled into a report and used to write a design brief that a design team should respond to, in order to address this problem. In 2011, working in partnership with the Department of Health, the Design Council launched a UK-wide open competition, aimed at finding a way to reduce the violence and aggression experienced daily by NHS hospital staff using design.

The winning design team was a UK-based multidisciplinary consortium, led by the design studio PearsonLloyd. Initially, the design team was supported by an independent advisory board, composed of senior stakeholders in health, industry, and education, organized by the Design Council to offer strategic guidance. The design team expanded the initial exploratory ethnographic research with their own field research to gain more focused insights into the problem and become familiar with the environment. This involved visits to NHS A&E departments to witness live incidents of aggression and violence, and watch staff in action (Figures 11.14 and 11.15).

Thanks to thorough analysis and interpretation of the collected data, the design team identified major areas of frustration and potential triggers of violence and aggression, in addition to those initially identified in the report. Main findings indicated that patients' confusion in the A&E environment and perceptions of not getting good care were key triggers for aggressive responses and violent attitudes. The lack of a human-centred approach also contributed to patients' poor experiences around A&E departments. Patients felt forgotten, neglected, and frustrated, responding with aggressive behaviour. Another set of findings indicated that patients didn't know how the A&E system worked; they had to go through a series of steps before getting help or treatment, but these steps weren't explicitly explained to them.

To shed light on these issues, the design team created a customer journey (Figure 11.16), to clearly articulate all steps involved in a patient's experience at the hospital: arrival, waiting, treatment, and final outcome (discharge or admission). Then, the design team focused on making each step a positive experience and keeping patients informed throughout their visit to A&E.

---

29 Understanding Violence and Aggression in Accident and Emergency Departments © ESRO 2011

Figure 11.14
Research outputs. Before generating designs, extensive research was carried out with designers, healthcare professionals, and psychologists to understand why incidents of violence and aggression occurred and where there were opportunities for improvement.

Figure 11.15
Research session. To ensure the designs were sensitive to the environment and beneficial to staff and patients, the team consulted with a cross-section of staff from partner hospitals.

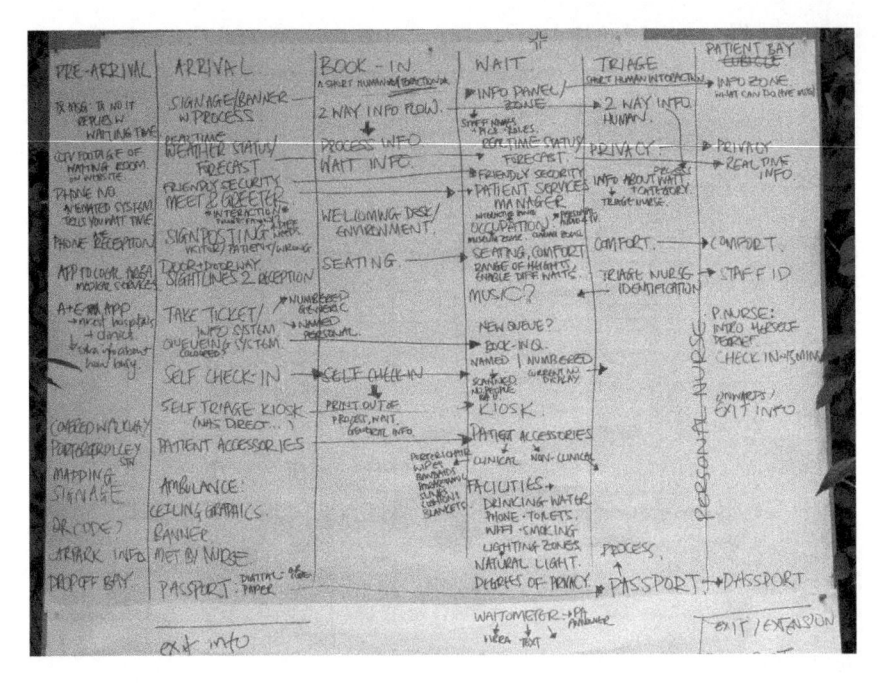

Figure 11.16
Research outputs. An important part of the project was mapping the patient journey through the emergency department from pre-arrival through to hospital discharge in order to understand what happens to patients at each stage.

The proposed solution for providing this needed guidance was to develop a 'slice' system, displayed as static, fixed signs[30]. A 'slice' would be inserted into any room, space, or corridor, becoming a communication point, helping patients know which step of the journey they were on. The design team also proposed training and discussing reflective practices with frontline staff, to learn tools to better manage patients' interactions and aggressive behaviours.

Formative evaluations: Computer models, mock-ups, and initial prototypes were created to test these ideas in real A&E departments and obtain feedback from staff and patients. Feedback on initial prototypes helped develop the concepts further and establish criteria to evaluate the final solution's performance. These had a flexible and adaptable design and layout, to allow compatibility and scalability with any A&E department across the country.

---

30 Reducing violence and aggression in A&E: Through a better experience. (Report, 2013) Design Council and Department of Health.

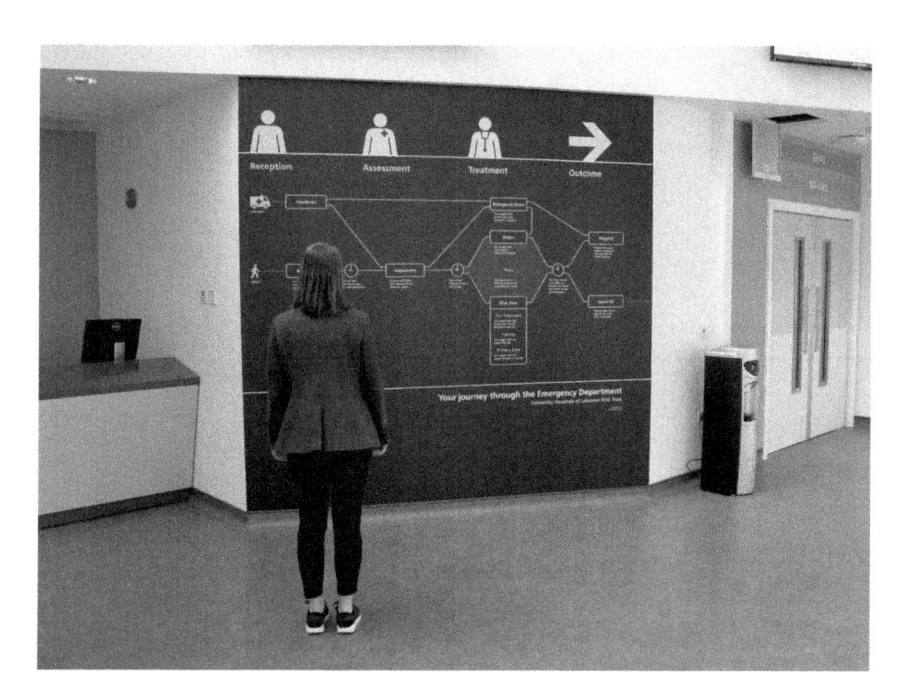

Figure 11.17
Final process map. A central portion of the final designs is the process map, which breaks the patient journey into four key stages and simply illustrates the treatment process with a series of stops—or waits—along the way. The process map is displayed in large format in the waiting room for all patients to see upon arrival.

Summative evaluations: In 2013, ESRO and Frontier Economics[31] were commissioned by the Design Council to conduct an impact evaluation study in order to assess the success of the new system. The evaluation combined patient and staff surveys with ethnographic observations and interviews. The slice system and the staff's programme were installed and tested in the A&E departments of two hospitals: Southampton General Hospital in Southampton and St George's Hospital in London. Two other hospitals of similar characteristics were selected as control groups: Oxford John Radcliffe Hospital in Oxford and King's College Hospital in London. Findings showed improved patient experiences, as the system made their treatment journey clearer, and a reduction in hostility and aggressive behaviour, because the new signage reduced their frustration during waiting times (Figure 11.17). Also, based on results from the patients' survey, the design team proposed extending the communications system and using screens to display live department waiting times.

31 Reducing violence and aggression in A&E: Through a better experience. An impact evaluation for the Design Council (Report, November 2013) © Frontier Economics Ltd, London.

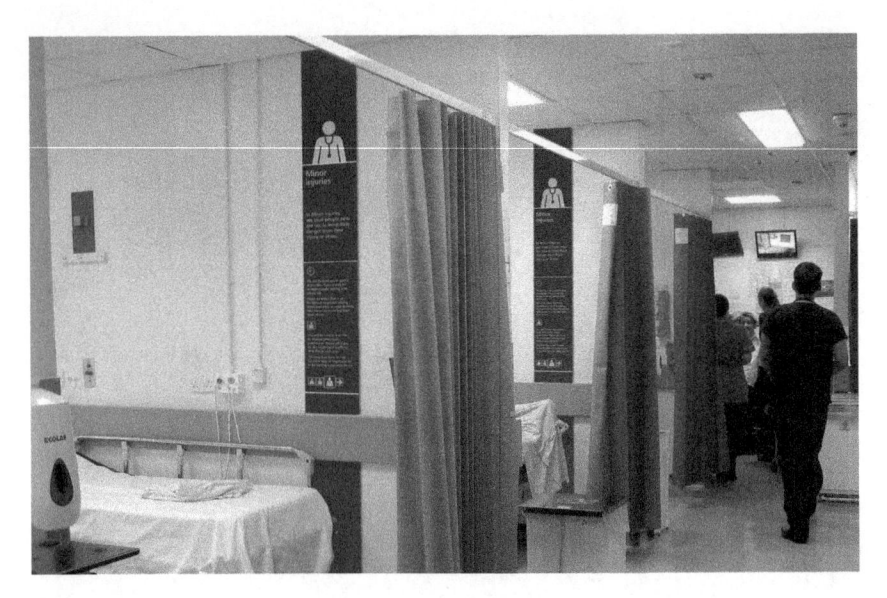

Figure 11.18
Final modular system. Panels are installed in each bay or room in which patients receive treatment. Information relevant to patients, such as where they are located, what type of staff will be treating them and where they are in the overall process, is displayed. Case study images from PearsonLloyd.

Solution

The final solution involved three outputs:

- **Guidance:** An on-site modular information and communication system, displaying key steps (slices) in respective rooms and A&E departments, to inform patients and reduce anxiety (Figure 11.18); a live digital screen, welcoming patients; and a large patients' journey wall map. Patient leaflets, live interactive media digital systems, and touch screen applications complemented the modular system, supporting the patient experience.

- **People:** A series of practical, people-centred training programmes, in-house inductions, or reflective sessions for staff members (trainee nurses, junior doctors, agency staff, receptionists, and security), to help them be better equipped to deal with incidents of violence and aggression.

- **Toolkit:** An illustrated publication and website for NHS frontline, management, estates, and industry, with recommendations to help improve patient experience and reduce violence and aggression.

The Design Council recommended the implementation of the new system in all NHS Trusts.

CASE STUDY 5:
To Park or Not to Park

New York, NY, US, 2014–Ongoing

Project team

Unlike the previous case studies, this project initially involved only the work of designer Nikki Sylianteng[32], later informed by parking judges, traffic engineers, police officers, a colour-blind council, and attention deficit disorder associations.

Project goal

The goal was to create a street parking sign that helped drivers to better understand parking rules and determine whether and for how long they could park in a specific parking spot.

Research approach and methods

- Desk research
- Non-participant observations
- Contextual interviews
- Virtual ethnography
- Surveys
- Formative and summative evaluations
- Participative design

Process

This project started as an experiment, in response to feelings of frustration and lack of understanding of street parking signs in the United States. After many parking tickets, in 2014, Sylianteng decided to redesign the signs by radically changing the approach and layout. Official street parking signs tend to vary from state to state, but they all often show more than two parking rules at the same time in the form of "sections": e.g. street sweeping

---

32 Official project website available at http://toparkornottopark.com/ [Accessed 10 November 2017].

Figure 11.19
Street parking sign in New Jersey. While some parking regulations may vary from state to state, most signs look similar.

schedule, street parking meter schedule, and exceptions (Figure 11.19). To gain a better sense of why official signs looked the way they did, Sylianteng interviewed traffic officials, parking judges, and other people associated with the design and use of parking signs. She learned that, up until then, the goal of official parking signs was to show valid parking rules, rather than to help drivers make a decision. In other words, signs show the rules or reasons someone could or could not park, but they don't explicitly indicate when drivers can or cannot park. In addition, while there are many rule changes over time, the signs aren't completely replaced each time there is a change. Instead, each rule change is added to existing signs as a new component. As a result, most signs include two, three, or more different sections, each representing a different rule that may or may not overlap with any of the previous ones. This piecemeal maintenance approach contributes to drivers' confusion and lack of understanding when trying to determine whether they can park there or not. In short, signs are designed based on how parking officials and engineers work, reflecting official language and terminology. For street parking signs to be clearer, parking rules should be simpler, so signs could use simpler language, and there should be a complete sign replacement each time there is a rule change[33]. However, this would be both expensive and time-consuming.

33 Gel Conference, 2015. Available at: https://vimeo.com/130161752 [Accessed 10 November 2017].

OLD   NEW

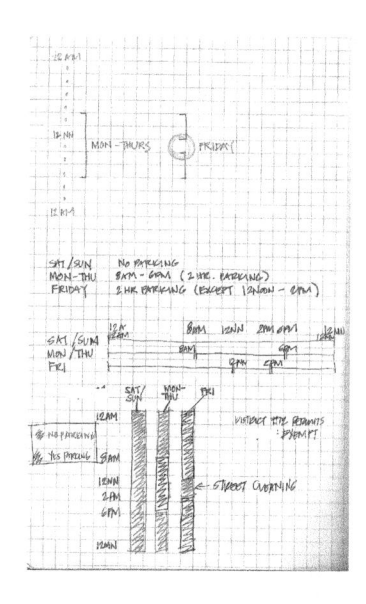

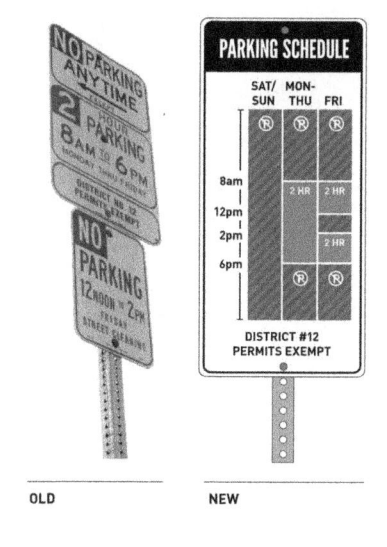

Figure 11.20

Process sketch. Early sketches breaking down the rules of an existing parking sign and exploring how to build it back up in a more straightforward way. The bottom half shows horizontal and vertical orientation explorations.

Figure 11.21

Old and proposed street parking signs. The timeline-based design simplifies the complicated process of elimination with existing parking signs. This way, the viewer arrives at the answer in two steps.

Based on these learnings, Sylianteng proposed redesigning official street parking signs from a user's perspective. The proposed parking sign focused on helping drivers make a decision and respond to two essential questions: 'Can I park here?' and 'For how long?'[34] but not on explaining each rule (the reason why). It also reduced the number of words and introduced a new visual language: a single 24-hour panel of a week that indicates when parking is allowed and for how long (Figures 11.20 and 11.21).

Formative evaluations: A change of this magnitude demanded a significant amount of testing. Initial studies focused on collecting users' opinions on the design concept and their level of understanding of the new visual language. A formative evaluation, using a low-cost prototype, was conducted in Brooklyn, NY, to assess whether people found the new sign easier to understand (Figure 11.22). The designer printed, laminated, and placed five to seven

34 ibid.

Figure 11.22
Formative evaluation study. The first prototype in Brooklyn, NY, used to test whether the general concept makes sense.

parking signs underneath official signs around the borough. People were invited to provide feedback by writing suggestions in a feedback box under the sign (a marker was attached, next to the sign) or online via Twitter or email.

One of the key findings from this study relates to the use of colours in the proposed sign. Colour-blind drivers found the initial combination of red and green frustrating. To address this issue, Sylianteng explored using different colour combinations, adding patterns, and adding icons to create a prototype that was readable and accessible to both colour-blind and non-colour-blind people. Following a participatory design approach, the designer collected insights from the public through social media, to inform design decisions. For example, she organized a colour-blind council to assess and provide feedback on different visual language versions of the sign. Through emails, people with attention deficit disorder contributed with suggestions, to find an appropriate balance in the use of icons, textual information, and colours, and traffic engineers provided ergonomic suggestions in relation to the implementation of the sign and the improvement of legibility from far away. Furthermore, people's comments helped identify the different parking situations that parking signs needed to address and determine how they could be scalable or adaptable to other cities with different or more parking rules (Figure 11.23).

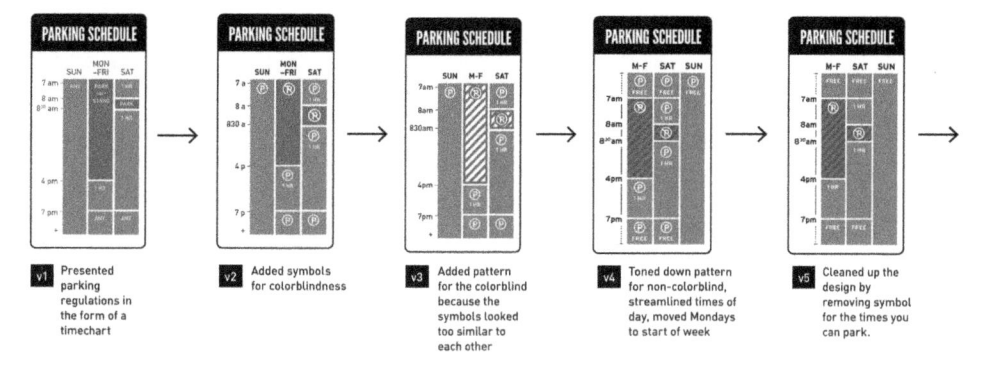

Figure 11.23
The evolution of the proposed new street parking sign.
Case study images from Nikki Sylianteng.

Summative evaluations: Thanks to the attention and visibility of the project in the press and social media, between 2014 and 2017, the new sign design concept was tested in various cities across the US, Canada, and Australia. In August 2014, the city of Vancouver (Canada) expressed an interest in doing a trial of the signs. A month later, staff from the city council in Los Angeles (US) requested permission to use the design. In 2015, Los Angeles performed a six-month trial, using the new signs as supplemental signage across seven city blocks in downtown. The study focused on testing the effectiveness and understandability of the new signs in context.

That same year, the parking sign concept was tested in Brisbane (Australia), during a three-month pilot in seven locations. This study used video-monitored observations and involved more than 600 contextual interviews to test the signs as supplemental signage, obtaining positive results.

For the first time, in another pilot conducted in New Haven (CT, US), signs were tested, replacing official signs rather than supplementing them. For six months, 100 foam-core parking sign prototypes were implemented. Finally, in March 2017, Oak Park village[35] also considered conducting a pilot to test the new signs.

Solution

A template to design a user-centred street parking sign.

35 Schering, S. (2017) Oak Park aims to consolidate parking signs through pilot study, *Chicago Tribune* [online], Available at: http://www.chicagotribune.com/suburbs/oak-park/news/ct-oak-parking-sign-pilot-tl-0323-20170317-story.html [Accessed 13 January 2018].

References

Anderson, K. (March, 2009) Ethnographic research: A key to strategy, *Harvard Business Review*, 87(3), 24.

Anderson K., Salvador, T., & Barnett, B. (2013) *Models in Motion: Ethnography Moves from Complicatedness to Complex Systems. EPIC 2013 Proceedings*, 1, 232–249.

Applied Information Group (2007) *Legible London. Yellow Book. A prototype wayfinding system for London*, London, UK.

Babbie, E. (2010) *The Basics of Social Research*, 5th edn, Wadsworth Publishing.

Beabes, M. & Flanders, A. (1995) Experiences with using contextual inquiry to design information, *Technical Communication*, third quarter, 409–420.

Bell, S.J. & Shank, J.D. (2007) *Academic Librarianship by Design: A Blended Librarian's Guide to the Tools and Techniques*, Chicago: American Library Association.

Bernard, H. & Ryan, G. (2010) *Analyzing Qualitative Data: Systematic Approaches*, Thousand Oaks, CA: SAGE.

Beyer, H. & Holtzblatt, K. (1998) *Contextual Design: Defining Customer-Centered Systems*, San Francisco: Elsevier.

Black, A., Luna, P., & Lund, O. (eds.) (2017) *Information Design: Research and Practice*, London: Routledge.

Blandford, A.E. (2013) *Semi-structured Qualitative Studies*, Interaction Design Foundation.

Blandford, A., Furniss, D., & Makri, S. (2016) Qualitative HCI research: Going behind the scenes, *Synthesis Lectures on Human-Centered Informatics*, 9(1), 1–115.

Bolger, N., Davis, A., & Rafaeli, E. (2003) Diary methods: Capturing life as it is lived, *Annual Review of Psychology*, 54(1), 579–616.

Bratsberg, H.M. (2012) *Empathy Maps of the FourSight Preferences*, Creative Studies Graduate Student Master's Projects, Paper 176.

Braun, V. & Clarke, V. (2013) *Successful Qualitative Research: A Practical Guide for Beginners*, Thousand Oaks, CA: SAGE.

Buzan, T. & Buzan, B. (1996) *The Mind Map Book: How to Use Radiant Thinking to Maximize Your Brain's Untapped Potential*, London: Penguin.

Chang, C. (2009) Making policy public: Vendor power! *Urban Omnibus* [online], Available at: http://urbanomnibus.net/2009/05/making-policy-public-vendor-power/ [Accessed 10 November 2017].

Chipchase, J. (2017) *The Field Study Handbook*, 2nd edn, Field Institute.

Clarke, V. & Braun, V. (2014) Thematic analysis, in: *Encyclopedia of Critical Psychology*, New York: Springer, 1947–1952.

Conley, C. (2004) Where are the design methodologists? *Visible Language*, 38(2), 196–215.

Conway, R., Masters, J., & Thorold, J. (2017) *From Design Thinking to Systems Change*, RSA, Action and Research Centre.

Cooper, A. (1997) *The Inmates are Running the Asylum: Why High Tech Products Drive us Crazy and How to Restore the Sanity*, Indiana: Sams Publishing.

Cross, N., Christiaans, H., & Dorst, K. (1996) *Analysing Design Activity*, Chichester: John Wiley & Sons.

Dahlbäck, N., Jönsson, A., & Ahrenberg, L. (1993) Wizard of Oz studies—why and how, *Knowledge-Based Systems*, 6(4), 258–266.

Design Council (2011) *Reducing Violence and Aggression in A & E: Through a Better Experience*, London, UK.

Dorst, K. & Lawson, B. (2009) *Design Expertise*, Oxford: Architectural Press.

Ericsson, K.A. & Simon, H.A. (1993) *Protocol Analysis*, Cambridge, MA: MIT Press.

Flick, U. (2009) *An Introduction to Qualitative Research*, 4th edn, Thousand Oaks, CA: SAGE.

Frascara, J. (ed.) (2015) *Information Design as Principled Action: Making Information Accessible, Relevant, Understandable, and Usable*, Champaign, Il: Common Ground Publishing.

Gage, N. (2012) *Making Emotional Connections Through Participatory Design* [online], Available at: http://boxesandarrows.com/making-emotional-connections-through-participatory-design/ [Accessed 19 November 2017].

Gaver, B., Dunne, T., & Pacenti, E. (1999) Design: Cultural probes, *Interactions*, 6(1), 21–29.

Given, L.M. (ed.) (2008) *The SAGE Encyclopedia of Qualitative Research Methods*, Thousand Oaks, CA: SAGE.

Gobert, I. & van Looveren, J. (2014) *Thoughts on Designing Information*, Zurich: Lars Müller Publishers.

Gray, D. (2017) *Updated Empathy Map Canvas* [online], Available at: https://medium.com/the-xplane-collection/updated-empathy-map-canvas-46df22df3c8a [Accessed 19 November 2017].

Gray, D., Brown, S., & Macanufo, J. (2010) *Gamestorming: A Playbook for Innovators, Rulebreakers, and Changemakers*, O'Reilly Media Inc.

Hall, E. (2017) Design Sprints are Snake Oil [online], Available at: https://medium.com/research-things/design-sprints-are-snake-oil-fd6f8e385a27 [Accessed 9 January 2018].

Hanna, P. (2012) Using internet technologies (such as Skype) as a research medium: A research note, *Qualitative Research*, 12(2), 239–242.

Heller, S. & Landers, R. (2014) *Infographic Designers' Sketchbooks*. New York: Princeton Architectural Press.

Henderson, S. & Segal, E.H. (2013) Visualizing qualitative data in evaluation research, *New Directions for Evaluation*, 2013(139), 53–71.

Holtzblatt, K. & Beyer, H. (2014), Contextual Design: Evolved, *Synthesis Lectures on Human-Centered Informatics*, 7(4), 1–91.

Irwin, T. (2002) Information design: What is it and who does it? *American Institute of Graphic Arts (AIGA)* [online], 21 June, Available at: http://online.sfsu.edu/jkv4edu/2DMG/projects/Informationdesign.pdf [Accessed 14 November 2017].

Jacobson, R. (2000) *Information Design*, London & Cambridge, MA: The MIT Press.

Jones, C.J. (1992) *Design Methods*, 2nd edn, New York: John Wiley.

Kawakita, J. (1986) *KJ hou*. Tokyo: Chuokoronsha.

Kelley, T. (2001) *The Art of Innovation*, Crown Business.

Kirk, A. (2012) *Data Visualization: A Successful Design Process*, Birmingham: Packt Publishing.

Klanten, R., Bourquin, N., & Ehmann, S. (2008) *Data Flow, Visualizing Information in Graphic Design*, Berlin: Gestalten.

Klein, G., Phillips, J.K., Rall, E.L., & Peluso, D.A. (2007) A data-frame theory of sensemaking, in *Expertise Out of Context, Proceedings of the Sixth International Conference on Naturalistic Decision Making*, Erlbaum, 113–155.

Knafo, S. (2009) Visual aids for the pushcart world, *The New York Times* [online], Available at: http://www.nytimes.com/2009/04/05/nyregion/thecity/05guid.html?_r=1&scp=5&sq=street%20vendor&st=cse [Accessed 13 January 2018].

Koskinen, I., Zimmerman, J., Binder, T., Redström, J., & Wensveen, S. (2011) *Design Research Through Practice: From the Lab, Field, and Showroom*, Waltham, MA: Morgan Kaufmann.

Ladner, S. (2014) *Practical Ethnography: A Guide to Doing Ethnography in the Private Sector*, Walnut Creek, CA: Left Coast Press.

Lawrence, N.W. (2013) *Social Research Methods: Qualitative and Quantitative Approaches*, 7th edn, Harlow: Pearson New International Edition.

Lawson, B. (2005) *How Designers Think. The Design Process Demystified*, London: Architectural Press.

Lincoln, Y.S. & Guba, E.G. (1985) *Naturalistic Inquiry*, Newbury Park, CA: SAGE.

Madrigal, D. (2009) *Contextual Interviews and Ethnography: Two Different Types of Home Visits* [online], Available at: http://usabilitypost.com/2009/09/09/contextual-interviews-and-ethnography/ [Accessed 14 November 2017].

Marcus, A. & Jean, J. (2009) Going green at home: The green machine, *Information Design Journal*, 17(3), 235–243.

Marshall, M.N. (1996) The key informant technique, *Family Practice International Journal*, 13(1), 92–97.

Mattelmäki, T. (2008) *Design Probes*, 2nd edn, Vaajakoski: University of Art and Design Helsinki.

Mattelmäki, T. & Battarbee, K. (2002) Empathy probes, in *PDC 02 Proceedings of the Participatory Design Conference*, 266–271.

McDonald, S. (2005) Studying actions in context: A qualitative shadowing method for organizational research, *Qualitative Research*, 5(4), 455–473.

McKim, R.H. (1980) *Experiences in Visual Thinking*, 2nd edn, Cengage Learning.

McQuaid, H.L., Goel, A., & McManus, M. (2003, June) When you can't talk to customers: Using storyboards and narratives to elicit empathy for users, in *Proceedings of the 2003 International Conference on Designing Pleasurable Products and Interfaces*, ACM, 120–125.

McQuaid, H.L., Goel, A., & McManus, M. (2003, September) Designing for a pervasive information environment: The importance of information architecture, in *Proceedings of the British HCI Conference*, Bath, UK.

Mika, M. (2016) The big(ger) picture: Why and how virtual ethnography can enhance generative design research, *Medium* [online], Available at: https://medium.com/sonicrim-stories-from-the-edge/the-big-ger-picture-why-and-how-virtual-ethnography-can-enhance-generative-design-research-c39acb92fe4e [Accessed 20 January 2018].

Miles, M. B., Huberman, A.M., & Saldaña, J. (2013) *Qualitative Data Analysis*, SAGE.

Morris, D. (1979) *Manwatching: A Field Guide to Human Behavior*, 3rd edn, Frogmore: Triad/Panther Books.

Nielsen, J. (1995) *10 Usability Heuristics for User Interface Design* [online], Available at: https://www.nngroup.com/articles/ten-usability-heuristics/ [Accessed 10 January 2018].

Nielsen, J. (1999) *Designing Web Usability: the Practice of Simplicity*, Indianapolis: New Riders Publishing.

Norman, D. (1988) *The Psychology of Everyday Things*, New York: Basic Books Inc.

Oppenheim, A.N. (1992) *Questionnaire Design, Interviewing and Attitude Measurement*, London: Continuum.

Patel, N. (2018) Everything is too complicated: What are you assuming people already know? *The Verge* [online], Available at: https://www.theverge.com/2018/1/7/16861056/ces-2018-bad-assumptions-smart-assistants-tech-confusion [Accessed 9 January 2018].

Patnaik, D. (2014) *Needfinding: Design Research and Planning*, Amazon.

Patton, M.Q. (2002) *Qualitative Research & Evaluation Methods*, 3rd edn, Thousand Oaks, CA: SAGE.

Petre, M. & Rugg, G. (2007) *A Gentle Guide to Research Methods*, Open University Press, McGraw-Hill.

Pettersson, R. (2010) Information design – Principles and guidelines, *Journal of Visual Literacy*, 29(2), 167–182.

Polanyi, M. (1983) *The Tacit Dimension*, Gloucester, MA: Peter Smith.

Pontis, S. (2012) *Guidelines for conceptual design to assist diagram creators in information design practice*, PhD thesis, University of the Arts London, UK.

Pontis, S. & Babwahsingh, M. (2013) Communicating complexity and simplicity: Rediscovering the fundamentals of information design, *2CO COmmunicating COmplexity*, Alghero, Sardinia, Italy, 25–26 October 2013, 244–261.

Pontis, S. & Babwahsingh, M. (2016) Start with the basics: Understanding before doing, in *VisionPlus 2015 Conference: Information+Design=Performance Proceedings*, (IIID, IDA, Birmingham, England), 90–102.

Pontis, S. & Babwahsingh, M. (2016) Improving information design practice: A closer look at conceptual design methods, *Information Design Journal*, 22(3), 249–265.

Pontis, S. & Blandford, A. (2015) Understanding "influence": An exploratory study of academics' process of knowledge construction through iterative and interactive information seeking, *Journal of the Association for Information Science and Technology*, 66(8), 1576–1593.

Pontis, S., Kefalidou, G., Blandford, A., Forth, J., Makri, S., Sharples, S., Wiggins, G., & Woods, M. (2016) Academics' responses to encountered information: context matters, *Journal of the Association for Information Science and Technology*, 67(8), 1883–1903.

Roam, D. (2013) *The Back of the Napkin: Solving Problems and Selling Ideas with Pictures*, London: Portfolio.

Roam, D. (2014) *Show and Tell: How Everybody Can Make Extraordinary Presentations*, London: Penguin.

Roberts, S. (2017) The UX-ification of research, *Stripe Partners* [online], Available at: http://www.stripepartners.com/the-ux-ification-of-research/ [Accessed 5 January 2018].

Rojas J. (2017) Etch A Sketch: How to Use Sketching in User Experience Design, *Interaction Design Foundation* [online], Available: https://www.interaction-design.org/literature/article/etch-a-sketch-how-to-use-sketching-in-user-experience-design [Accessed on 19 December 2017].

Rose, G. (2012) *Visual Methodologies: An Introduction to Researching with Visual Material*, Thousand Oaks, CA: SAGE.

Rubin, J. & Chisnell, D. (2008) *Handbook of Usability Testing: How to Plan, Design, and Conduct Effective Tests*, Chichester: John Wiley & Sons.

Russell, K. (2002) Why the culture of academic rigour matters to design research: Or putting your foot into the same mouth twice, *Working Papers in Art and Design*, 2 [online], Available at: https://www.herts.ac.uk/__data/assets/pdf_file/0007/12310/WPIAAD_vol2_russell.pdf [Accessed 17 January 2018].

Saldaña, J. (2015) *The Coding Manual for Qualitative Researchers*, Thousand Oaks, CA: SAGE.

Sanders, E.B.N. (2002) From user-centered to participatory design approaches, Chapter 1 in Frascara, J. (ed.) *Design and the Social Sciences: Making Connections*, London: Routledge.

Sanders, E.B.N. (2016) Design research in 2006, *Design Research Quarterly*, V. I(1), September, 1–9.

Sanders, E.B.N. & Stappers, P.J. (2008) Co-creation and the new landscapes of design, *International Journal of CoCreation in Design and the Arts*, 4(1), 5–18.

Sanders, E.B.N. & Stappers, P.J. (2014) Probes, toolkits and prototypes: Three approaches to making in codesigning, *CoDesign*, 10(1), 5–14.

Sangasubana, N. (2011) How to conduct ethnographic research, *The Qualitative Report*, 16(2), 567–573.

Schering, S. (2017) Oak Park aims to consolidate parking signs through pilot study, *Chicago Tribune* [online], Available at: http://www.chicagotribune.com/suburbs/oak-park/news/ct-oak-parking-sign-pilot-tl-0323-20170317-story.html [Accessed 13 January 2018].

Schriver, K.A. (1997) *Dynamics in Document Design: Creating Text for Readers*, New York: Wiley.

Siegel, A. & Etzkorn, I. (2013) *Simple: Conquering the Crisis of Complexity*, London: Twelve.

Sleeswijk Visser, F. (2009) *Bringing the everyday life of people into design*, PhD thesis, Delft University.

Sleeswijk Visser, F., Van der Lugt, R., & Stappers, P.J. (2004) The personal cardset—a designer-centred tool for sharing insights from user studies, in *Proceedings of Second International Conference on Appliance Design*, Bristol, 157–158.

Sless, D. (2008) Measuring information design, *Information Design Journal*, 16(3), 250–258.

Sless, D. (2012) Design or "design" – Envisioning a future design education, *Visible Language*, 46(1/2), 54.

Sommer, B. (2006) *Participant observation*, Department of Psychology, University of California [online], Available at: http://psc.dss.ucdavis.edu/sommerb/sommerdemo/observation/partic.htm [Accessed 13 January 2018].

Steven, R. (2013) A better A&E [online], Available at: https://www.creativereview.co.uk/a-better-ae/ [Accessed 10 November 2017].

Sylianteng, N. (2015) To Park Or Not To Park, *Gel Conference* [online], Available at: https://vimeo.com/130161752 [Accessed 10 November 2017].

Treffinger, D.J., Isaksen, S.G., & Stead-Dorval, K.B. (2005) *Creative Problem Solving: An Introduction*, 4th edn, Waco, TX: Prufrock Press Inc.

Ulrich, K.T. & Eppinger, S.D. (2012) *Product Design and Development*, 5th edn, London: Irwin McGraw-Hill.

Valsiner, J. & Rosa, A. (2007) *The Cambridge Handbook of Sociocultural Psychology*, Cambridge: Cambridge University Press.

Visser, F.S., Stappers, P.J., Van der Lugt, R., & Sanders, E.B.N. (2005) Contextmapping: Experiences from practice, *CoDesign: International Journal of CoCreation in Design and the Arts*, 1(2), 119–149.

Waller, R. (2011) *Technical Paper 14, Information design: How the disciplines work together*, Simplification Centre, University of Reading.

Ware, C. (2008) *Visual Thinking for Design*, San Francisco: Elsevier.

Weick, K.E. (1995) *Sensemaking in Organizations*, Thousand Oaks, CA: SAGE.

Weinstein, L. (2015) *An Inside Look at Design Tools in Development Work: User Personas in Context* [online], Available at: https://reboot.org/2015/06/18/user-personas-in-context/ [Accessed 31 August 2017].

Whiteside, J., Bennett, J., & Holtzblatt, K. (1988) Usability engineering: Our experience and evolution, in Helander, M. (ed.) *Handbook of Human Computer Interaction*, Amsterdam: Elsevier.

Willig, C. (2013) *Introducing Qualitative Research in Psychology*, London: McGraw-Hill Education.

Wood, J. (2000) The culture of academic rigour: Does design research really need it?, *The Design Journal*, 3(1), 44–57(14).

Wurman, R.S. (1989) *Information Anxiety*, New York: Doubleday.

Wurman, R.S. (1996) *Information Architects*, Zurich: Graphics Press.

Index

Page numbers in **bold** refer to figures, page numbers in *italic* refer to tables.

A

A&E project *204*, 225–6, **227**, 228–30, **228**, **229**, **230**
abbreviation system 73
accuracy 45
actionable items, translating findings into 183–96, **183**, **185**
action items 191
action research **18**, 21–2
actions *116–17*
activities 106–8
activity needs 10, **11**
Adobe Bridge 146
affinity diagrams 140, **148**, 158
agile processes xvi
ambiguity 38
analysis xvii, 32, 130; unit of 59–60
analysis output **209**; *see also* data analysis and interpretation
Anderson, Ken 19
anthropology 12, 17, 19
applicability 44, 47, 176–7
applied ethnography 19
applied research **32**
Applied Wayfinding 9, 211–15, **213**, **214**, **215**, **216**, 217–18, **218**
artefacts 97
assumptions 197; working with xv–xvi
asynchronous online interview 112
attitudes, participants 63
audiences 4, 11; basic knowledge xv–xvi; cognitive activities 4, 113; context 9; empathy 174; environment 9; findings 171, 172; identification 35; initial responses 33; lifestyles 9; needs 31, 193–4; segmenting 160; understanding 9–10, 31–2
audio recording 78
authenticity 162, 173–4
authority 220, 222

B

baseline knowledge 48
basic knowledge, audiences xv–xvi
before- and after-design 118–19
behavioural changes 114, *115*
behaviours, and cultural backgrounds 39–40
biases 40, 131
Big Q studies 24, **25**, 29
binary questions 91
Black, A. et al, *Information Design: Research and Practice* 7
body language 77
brainstorming 187, **206**; visual 191–2, **192**
budget 48–9

C

calendar 69, **70**
cardsets 178–9, **178**
Carnegie Library of Pittsburgh redesign project 193, *204*, 205–8, **206**, **209**, 210, **210**
categories 140, **142**
category creation 149
causes 141
Centre for Urban Pedagogy 219–20
checklists 71
chronologies 144
clients, role 27
coded images 150
codes and coding 73, 130; affinity diagrams 152; combining **142**; definitions 137, **138**; framework 135; generating 135, 137, 139; labels 137, **138**; number of 139; open coding 135, **136**; revision 137, *139*; roadmaps 135; saturation 139; support 137; visual content analysis 149–50
codesign 21
cognitive activities, audiences 4
cognitive psychology 21
cognitive walkthroughs 118
collaborative workshops *82–3*, 104–9, 106–8,

Index | 245

107, **110**, 122, 124, 205–8, 210, 225–6, 228–30
collage activities 106–7
common needs 10, **11**
communication 131, 171; dimensions of 171–2, **173**; techniques 172, **173**
compensation 43
complexity 4
concept design 32–3
concept evaluations 122
concept maps 208
concepts: evolving 194–5; grouping 130; validation 195
conceptual design 30, 31–3
conceptual framework 56
conclusions 176–7
confidentiality 42–3
confirmability 44, 47
connections: making 4; visualizing 143–4
connections and relationships visualizations 164–5, **165**
consequences 141
consistency 47
constraints 59; working around 48–9
constructive-interpretatives 27, 28–9, **28**
consultation study 215, **215**
content 13
content expertise 48
context 21, 97, 206, 208; audiences 9
context needs 10, **11**
contextual inquiry **18**, 20, 82–3, 95–7, **96**, **98**, 124, **181**
contextual interviews 82–3, 86, 89–93, **91**, **94**, 122, 205–8, 210, 219–20, 222, 224, 231–5
contradictions 144
convenience sampling 61
core in-depth questions 93
covert evaluations 123–4
covert observation 84
creativity, field research 49–50
credibility 44, 45–6, *46*, 131, 172, 176
cross-cultural participants 41
cues 154
cultural context 39, 39–40
cultural knowledge 39–40
cultural probes 22
cultural trends 9
culture: and methods 40–1; role in field research 39–41
customer experiences 19

D

data access 73
data analysis and interpretation 55, 66, 129–67; affinity diagrams **148**, 151–2, **153**; analysis 130; Carnegie Library of Pittsburgh redesign 208, **209**; category creation 140, **141**, **142**; checking 132; coded images 150; coding 130, 135, **136**, 137, **138**, 139, *139*, **142**; communication 131; completeness 132; connection visualization 143–4; core tasks 129; credibility 45; data preparation 132, 134–5; data quality 134; empathy maps **148**; Five Ws + One H 147–8, **148**; formatting 134; framework 131; insights 129; interpretation 130–1; interpretation methods 157–67, **157**; needfinding 157–8, **157**, *159*; organizing 147–56; page-by-page analysis 135; personas 154, **157**, 159–60, *161*, 162, **162**; process 132, **133**, 134–44, **136**, **138**, *139*, **141**, **142**, **145**; qualitative research 26; sensemaking 129–32, **133**, 134–5; set up 134–5; sorting 132; story creation 144; synthesis 131; theme identification 140–3; tools 145–7, **145**; visual content analysis **148**; visualizations **157**, 163–7, **164**, **165**, **167**, **168**
data collection 55, 63–6, 126, 197; collaborative workshops 109; concept evaluations 122; contextual inquiry 96–7; contextual interviews 92–3; covert evaluations 124; design probes 103–4; diary studies 103–4; free evaluations 128; observational studies 87–8, *88*
data evaluation 103–4
data, proliferation of xv
data recording 72–3
data visualizations 5, 16, 50, 52, **157**, 163; connections and relationships 164–5, **165**; evolution and sequences 165–7, **167**, **168**; portraits 163–4, **164**
datasets, analysis & synthesis 32
debriefing 104
decision-making 9, 22
deductive analysis 37n1
dependability 44, 47
descriptions 175–6; thick 46
descriptive labels 137, **138**
descriptive observations 87
design briefs xvi
design concepts 192–3

Design Council UK 225–6, **227**, 228–30, **228**, **229**, **230**
design decisions 4, 183–96
design discussions **216**
design experience 9
design probes 22–3, *82–3*, **105**, 126; package 99, 99–100, **101**, 102
design process stages **194–5**, **198–9**; early 193–4; findings 171–2; late 195–6; middle 194–5
design research 20, **25**
design researchers **85**, **91**, **96**, **100**, **107**, 184, 197, **198**
design sprints xvi
design study 57, 95, **198–9**
design technology, availability of xv
desk research 60, 211–15, 217–18, 219–20, 222, 224, 225–6, 228–30, 231–5
detail design 33–4
diagrams 164–5, **165**, 166, 167
diaries 78
diary studies 22, 23, *82–3*, **105**, 126, 127–8
digital format 99
digital self-documentation studies 111
digital tools 146–7
direct-experience storyboards 205–8, 210
direction, defining 193
discussion, sparking 182
document design 5
Dunne, Tony 22
duration 86–7
dynamic information environments 5

E

educational resources 219–20, **221**, 222, **222**, **223**, 224
eligibility criteria 61, *62*
emotions 78, 106
empathic focus 9–10
empathy 22–3, 37, 162, 174
empathy maps **148**, 154–6, **155**, *156*
environment, audiences 9
ergonomics 21
ethics and ethical issues 42, 85–6
Ethnographic Praxis in Industry (EPIC) international conference 19
ethnography 17–20, **18**, 225–6, 228–30, 231–5
evaluation reports 66
evaluation studies 34, 113; concept evaluations 122; covert 123–4; dimensions 113–15, *116–17*; field evaluations 118–21; formative 115, 118, 123–4, 205–8, 210, 219–20, 222, 224, 225–6, 228–30, 231–5, **234**; free 126–8; overt 124–6; qualitative measure 120; summative 34, 115, 118, 217–18, 225–6, 228–30, 231–5; types of 120–1, 121
evaluative questions 115, *116–17*
evaluative studies 35–6
evidence, chain of 175
evolution and sequences visualizations 165–7, **167**, **168**
expert consultations 60
expertise 48
Experts Persona 193–4
explicit findings 183
explicit knowledge 64, **67**
explicit needs **67**
exploratory studies 34–5, 81, 212; collaborative workshops 104–9, 107, **110**; contextual inquiry *82–3*, 95–7, **96**, *98*; contextual interviews *82–3*, 89–93, **91**, *94*; design probes *82–3*, 97–100, **99**, **101**, 102–4, **105**; diary studies *82–3*, 97–100, **99**, **101**, 102–4, **105**; observational studies 81, *82–3*, 83–8, **85**, *88*, *89*; online field research 109, 111–12
external validation 46–7

F

Facebook 111
feasibility testing 123, 217
feedback 9, 214, 222, 224
feelings 78
Fendley, Tim 9
field evaluations 118–21
field interviews 211–15, 217–18
field notes 76–8, **77**, 78–9, **79**
field observations 20
field research: ambiguity 38; budget 48–9; communication 40; conducting 38–43, **39**; constraints 48–9; creativity 49–50; cultural context 39; definition xvii; empathy 37; ethics 42–3; evaluative 35–6; exploratory 34–5; and information design 12–13, **13**, 17–36; information design aids 50, **51**, 52; interpretation 37–8; key components 37–8; methods xvii–xviii; methods and culture 40–1; natural settings 37; origins 12; process 55, **57**; quality and validity criteria 44–7, **45**, *46*; questions **14**, **15**; rapport

Index | 247

building 41; reflexivity 38–9, 40; rigour 43; role of culture 39–41; sensibility 44; skills xviii, 44; timing 34–6; use of xvii; working with 29–34, **32**, **33**

field studies 55–80, **57**, 197; calendar 69, **70**; checklist 80; constraints 59; data analysis 66; data collection 63–6, **67**; design 55–6, **57**, **58**, 59–66, **62**, **67**, 68; design activities 56; desk research 60; direction 55–6; duration 59; expert consultations 60; field sessions 76–9, **77**, **79**; findings dissemination 66, 68; goal setting 56; literature review 60; motivation 55; participants 59, 61–3, **62**; piloting 75; resources 59; specific focus 59–60; study roadmap 68–9, **69**; teams 68; testing 75; timeline 69, **70**; toolkit 70–3, **71**, **72**, **73**, **74**, 75; unit of analysis 59–60; workspace 75

field visits 95

finding cards 186, 186–7, **187**

findings: explicit 183; generative activities 187; interpretation 184, 186; organizing 186–7; transformation approaches 186–92, **188**, **190**, **192**; translating into actionable items 183–96, **183**, **185**

findings, reporting 171–82, 208; audiences 171, 172; communication dimensions 171–2, **173**; conclusions 176–7; descriptions 175–6; design process stage 171–2; format 172; goals 173; interpretation 175–6; key parts 177–80, **177**, **178**, **181**, **182**; main storyline 175–6; personal cardsets 178–9, **178**; personas 179; scenarios 179–80; story 173–4; storyboards 180, **181**, **182**; study overview 175; themes 175; unfiltered descriptions 174; validation 176; visual material 174; whole study 174–7

Five Ws and One H 147–8, **148**, 189

flowcharts 165

focus 59–60, 86, 90, 95, 122

focused observations 88

formative evaluations 115, 118, 123–4, 205–8, 210, 219–20, 222, 224, 225–6, 228–30, 231–5, **234**

formative field evaluations 33, 33–4

Four Quadrants chart 188–9, **188**

framed problems 6

Frascara, J., *Information Design as Principled Action* 7

free evaluations 126–8

frequencies 144

functionality tests 211–15, 217–18

G

Gaver, Bill 22

generality, level of 10, 158; needs 10, **11**

general needs 9

generative activities 106, 109, 187

goals 86, 93, 95, 99, 104, 122, 123, 126, 173

Google Ventures xvi

Gray, D. 154n18

guides 71, **72**, 92

guiding questions **25**

H

heuristic evaluations 118

hierarchical structures 165

hierarchies 144

human-centred approach 226

human computer interaction 20

I

icons 5

ideas: evolving 194–5; generation 194; grouping 130; unexpected 137; validation 195

identified needs 158

IDEO Company 19

ideographs 164, 166

immersive approach 9

implementation 34, 208

implicit knowledge 64, **67**

incidents analysis 228–30

in-depth interviews 20

inductive analysis 37–8

inferences 78

informal studies 113

information design: cross-disciplinary nature 3; definition 3; dimensions 4, **5**; effective xvi, 4, **5**; and field research 12–13, **13**, 17–36; problem facing xv–xvi; rise in interest in xiii; role of people 7, 9–10; skills xviii

information design aids, field research 50, **51**, 52

information design challenges 5–7

information designers: challenges xv; core tasks xix; myths and misconceptions xvii; role of 7, 9, 11–12; skills 3, 11, 13, 16; specializations 5–6

information design process 30n33;

248 | Index

analysis and synthesis 32; audience understanding 31–2; detail design 33–4; evaluation 34; field research timing 34–6; implementation 34; problem understanding 31; prototype design 33–4; research-led 29–34, **32**, **33**; subject matter understanding 31–2

information design researchers 14, 16, **85**, **91**, **96**, **100**, **107**, 184, 197, **198**

information design techniques 12

information-intense situations 20

informed consent 87

informed consent forms 42

initial interpretations 78

initial interviews 103, 126

insights 129

inspiration 194

instructions 102, 108

instruments 70, 108

interactive projects 21

internal validation 45

Internet, the xv, 109, 111–12

interpretation 37–8, 130–1, 135n9. *see also* data analysis and interpretation

interpretative lens 130

intervention, opportunities for 196

interview guide **72**, 92

interviews xix; affinity diagrams **153**; asynchronous 112; contextual *82–3*, 86, 89–93, **91**, **94**, 122, 219–20, 222, 224, 231–5; data collection 92; empathy maps **155**; field 211–15, 217–18; final 104; guides **72**, 92; in-depth 20; initial 96, 103, 126; online 111, 112; participants 92; questions 90, 90–1, 92–3; semi-structured 90; synchronous 112

investigative rigour xvii

Irwin, Terry 3

J

jargon 172

journeys 166–7, **167**, **168**, 208, **209**, 225–6, 228, 228–30, **228**, 229

K

Kawakita, Jiro 151n17

key concepts **198–9**

key informants 90, 211–15, 217–18

key words 71, 76

KJ method 151n17

knowledge, types of 64, **67**

L

labels 137, **138**

languages 220

latent needs **67**, 85, 95, 104, 109, 214

leading questions 91

Legible London project *204*, 211–15, **213**, **214**, **215**, **216**, 217–18, **218**

lifestyles, audiences 9

listening 37, 93

literature review 60

Livingstone, Ken 212

location 108

logs 72, **74**

London: Congestion Charging Zone 212–13; Legible London project 211–15, **213**, **214**, **215**, **216**, 217–18, **218**

London Underground map **151**

looking 87

Lovejoy, Tracey 19

M

magic points 166

magnitudes 140

maps 6, 9

market research 9

materials 70, **71**, 108

Mattelmäki, Tuuli 22–3

MAYA 205–8, **206**, **209**, 210, **210**

meaning: finding 29; generating 130–1, 143–4

memory, types of 64

memos 78

mental mapping 211–15, **214**, 217–18

metrics 217

mind maps 56, **58**, 140

mixed studies 36

monitoring 103

mood 78

motivation, field studies 55

myths and misconceptions 197

N

National Health Services 225–6, **227**, 228–30, **228**, **229**, **230**

natural settings 37

needfinding 157–8, **157**, *159*

needs 99; audiences 31, 193–4; categories of 10, **11**; explicit needs **67**; latent **67**, 95, 104, 109, 194 level of generality 10, **11**

need statements 158, *159*

negative cases 46

Index | 249

Nielsen, Jacob 21
non-participant observation *82–3*, 84, **85**, 123, 225–6, 228–30, 231–5
non-verbal cues 154
Norman, Don 21
note-taking 73, 76–8, **77**, **206**

O

objective measures 23
objects and instruments, self-documentation studies 99–100
observation guide 88, **89**
observational studies 81, 84–8, **85**, **88**, **89**, 122, 123–4, 205–8, 210, 211–15, 217–18, 219–20, 222, 224, 225–6, 228–30, 231–5
online collaborative sessions, 111
online field research 109, 111–12
online forums 111
online interviewing 111, 112
online polls 214
open coding 135, **136**
open discussion 109
open-ended questions 91
openness 49
open-questions 23–4
opportunistic sampling 62
optimization 195–6
outcomes 114, *116–17*
overt evaluations 124–6
ownership 214

P

Pacenti, Elena 22
page-by-page analysis 135
pain points 166, 183, 196
paradigms 27
parking sign project *204*, 231–5, **232**, **233**, **234**, **235**
participant observation 17–8, *82–3*, 84–5, **85**, 206, **207**, 208
participants: attitudes 63; collaborative workshops 104, 106, 108; concept evaluations 122; contextual inquiry 96, **96**; contextual interviews **91**, 92; covert evaluations 124; design interaction 119; design probes 97–8, **100**, 102–3, 103–4; diary studies 97–8, **100**, 102–3, 103–4; eligibility criteria 61, *62*; field notes 76–7; field studies 59, 61–3, *62*; free evaluations 126, 128; observational studies 87; online interviewing 112; overt evaluations 124, 125; quotes 137, 143, 162; raw data 178–9; recruitment 61, *62*, 63, 96, 197; reflective time 103–4; sample size 61, *62*; sampling strategy 61, 61–2, *62*
participative design 231–5
participatory action research **18**, 21–2
participatory design 21
partnership 97
PearsonLloyd 225–6, **227**, 228–30, **228**, **229**, **230**
people-centred 7, 9–10, 17–23, **18**, 29
performance evaluation 113–15
personal cardsets 178–9, **178**
personal views 45–6
personas 154, **157**, 159–60, *161*, 162, **162**, 179, 193–4
philosophical traditions 27
photoboarding 182
pictographs 164, 166
piloting 75
pilot studies 75, 92, 102, 217–18
points of view 27
portraits 163–4, **164**
positivists 27–8, **28**
practice, and research 16
precision 49
preparation: collaborative workshops 106–8; concept evaluations 122; contextual inquiry 95; contextual interviews 90–2; covert evaluations 123; design probes 99–100, 101; diary studies 99–100, 101; free evaluations 127; observational studies 86–7; overt evaluations 125
presentations 174
primary data 60
prior work 46
privacy 42–3
privacy concerns 206
problems: ambiguous 7; boundaries 7; focus on 13; framed 6; framing 34–5; spectrum of **8**; understanding 31, 206, 208; unframed 6–7
process diagrams 166, *167*
processes 141, 144
process maps **229**
process sketches **233**
prototype design 30, 33–4
prototypes, testing 217
public transport information 211–15, **213**, **214**,

250 | Index

215, **216**, 217–18, **218**
purposeful sampling 62

Q
qualifier needs 10, **11**
qualitative data analysis 66
Qualitative Data Analysis tools 146–7
qualitative data collection 63
qualitative paradigm 29
qualitative research xv, xvii, 16, 23, 23–9, 36; Big Q studies 24, **25**, 29; client role 27; criticisms 26; data analysis 26; lack of trust in 24, 26–9, **28**; methods **25**; paradigms 27, 44; role xix
quality and validity criteria 44–7, **45**, *46*
quantitative research 23, 37n1, 38, 44
quantitative studies 211–15, 217–18
questions: binary 91; core in-depth 93; empathy maps 155, *156*; evaluative 115, *116–17*; generating 90–1; guiding **25**; interviews 90, 90–1, 92–3; leading 91; open 23–4, 91; research 56; warm-up 92; wording 91
quotes, participants 137, 143, 162

R
rapid prototyping activities 107–8
rapport 41
raw data 174; participants 178–9
reactions 78
recruitment 61, *62*, 63
Reddit 111
refinement 208
reflective time, participants 103–4
reflexivity 38–9, 40
reminders 103
reports 174
research 9, 12; and practice 16
research data 12, 135, 174, **213**, **214**; *see also* data
research emphasis 29–30
research expertise 48
research methods xvi
research outputs **206**, **207**, **227**, **228**
research process xviii, xix, xx, 16, 26, 36, 50, 52, 55, **57**, 66, 68, 75, 80, 106, 163, 171, 174, 180, **185**, 197
research questions 56
resources 48–9, *49*, 59, **198–9**
rigour xvii, 43, 49, 103, 129

S
safety 43
sample size 61, *62*, 87, 92, 96
sampling strategy 61, 61–2, *62*
satisfaction 114, 115, *116–17*
satisfaction levels 218
saturation, codes and coding 139
scenarios 179–80
seeing 87
segmenting 160
selective observations 88
self-documentation **18**, 22–3, 97–100, 111, 127
self-documentation package 99–100, **101**, 102
semi-structured interviews 90
Sense Information Design project 149, 149–50
sensemaking xix; activities 129; analysis 130; communication 131; definition 4; interpretation 130–1; process 132, **133**, 134–44, **136**, **138**, *139*, **141**, **142**, **145**; visualizations 163
shadowing *82–3*, 84, **85**, 95
show & tell 109
signs 6, 231–5, **232**, **233**, **234**, **235**
situations, processing 4
Sless, David xvii
small q studies 24
snowball sampling 62
sociology 12
specializations, information designers 5–6
stakeholders 21, 208, 215
story 144, 173–4
storyboard activities 107
storyboards 180, **181**, 182; direct-experience 205–8, 210
structures 141
studio rigour 43
study overview 175
study roadmap 68–9, **69**
subjective design 10
subject matter, understanding 31–2
summative evaluations 34, 115, 118, 217–18, 225–6, 228–30, 231–5
support themes 143
surveys 90, 231–5
Sylianteng, Nikki 231–5, **233**, **234**, **235**
symbols 5
synchronous online interview 112
synthesis 32, 131
systems 6–7, **8**, 12, **14**, 123, **204**, 212

Index | 251

T

tacit knowledge 64–5
tasks 99; self-documentation studies 99, 100, 102
teams 11, 68
templates 71, **73**, **74**; empathy maps 154; journeys **167**, **168**; personal cardsets **178**; personas 160
test guides 125
testing, field studies 75
testing sessions 222, 224, 233–4, **234**
textual narrative 172
themes: connection visualization 143–4; identification 140–3, 152; names 143; number of 143; refining 143; reporting 175; supporting 143
thick descriptions 46
tiger teams 208, **210**
time frames 48, **198–9**
timelines 69, **70**, 165–6
Today pictures 189–91, **190**
Tomorrow pictures 189–91, **190**
tools 72–3; data analysis and interpretation 145–7, **145**
touchpoints 166, 196
transferability 44, 46–7
trends 140
triangulation 45, 46, 176
Twitter 111

U

uncertainty, tolerating 38
understanding 114–15
unexpected, the 49–50, 137
unfiltered descriptions 174
unframed problems 6–7
unit of analysis 59–60
universal needs 10
usability and usability testing **18**, 21, 24, 114, 115, *116–17*, 119, 124
user-centred approach 19
user experience cycles **209**
user experience (UX) design 21
user research xvi, 154

V

validation 176; concepts 195
validity criteria 44–7, **45**, 46
Vendor Power Guide project 204, 219–20, **221**, 222, **222**, **223**, 224
verbal cues 154
verbs 158
video recording 78
virtual data 111
virtual ethnography 19–20, 111, 231–5
virtual spaces 19–20
visual brainstorming 191–2, **192**
visual content analysis **148**, 149–50, **150**, **151**, 211–15, 217–18
visual displays 50
visual explanations 5
visualizations 5, 16, 50, 52, **157**, 163; connections and relationships 164–5, **165**; evolution and sequences 165–7, **167**, **168**; portraits 163–4, **164**
visual material 174

W

Ware, Chris 222
warm-up questions 92
wayfinding, Legible London project 211–15, **213**, **214**, **215**, **216**, 217–18, **218**
wayfinding study 213–14, **214**
Wizard of Oz technique 127–8
Wood, J. 43
workflow 165
workshops 22
workspace 75